Seismic Analysis and Design using the Endurance Time Method

Taylor and Francis Series in Resilience and Sustainability in Civil, Mechanical, Aerospace and Manufacturing Engineering Systems

Series Editor
Mohammad Noori
Cal Poly San Luis Obispo

Published Titles

Resilience of Critical Infrastructure Systems
Emerging Developments and Future Challenges
Qihao Weng

Experimental Vibration Analysis for Civil Structures
Testing, Sensing, Monitoring, and Control
Jian Zhang, Zhishen Wu, Mohammad Noori, and Yong Li

Reliability and Safety of Cable-Supported Bridges
Naiwei Lu, Yang Liu, and Mohammad Noori

Reliability-Based Analysis and Design of Structures and Infrastructure
Ehsan Noroozinejad Farsangi, Mohammad Noori, Paolo Gardoni, Izuru Takewaki, Humberto Varum, Aleksandra Bogdanovic

Seismic Analysis and Design using the Endurance Time Method
Edited by Homayoon E. Estekanchi and Hassan A. Vafai

Thermal and Structural Electronic Packaging Analysis for Space and Extreme Environments
Juan Cepeda-Rizo, Jeremiah Gayle, and Joshua Ravich

For more information about this series, please visit: https://www.routledge.com/ Resilience-and-Sustainability-in-Civil-Mechanical-Aerospace-and-Manufacturing/ book-series/ENG

Seismic Analysis and Design using the Endurance Time Method

Second Edition

Edited by
Homayoon E. Estekanchi and Hassan A. Vafai

CRC Press
Taylor & Francis Group
Boca Raton London New York

CRC Press is an imprint of the
Taylor & Francis Group, an **informa** business

Second edition published 2022
by CRC Press
6000 Broken Sound Parkway NW, Suite 300, Boca Raton, FL 33487-2742

and by CRC Press
2 Park Square, Milton Park, Abingdon, Oxon OX14 4RN

© 2022 selection and editorial matter, Homayoon E. Estekanchi and Hassan A. Vafai; individual chapters, the contributors

First edition published by Momentum Press 2018.

CRC Press is an imprint of Taylor & Francis Group, LLC

Library of Congress Cataloging-in-Publication Data
Names: Estekanchi, Homayoon E., editor. | Vafai, Hassan A., editor.
Title: Seismic analysis and design using the endurance time method /
edited by Homayoon E. Estekanchi and Hassan A. Vafai.
Description: Second edition. | Boca Raton, FL: CRC Press, 2022. |
Series: Resilience and sustainability | Includes index.
Identifiers: LCCN 2021020844 (print) | LCCN 2021020845 (ebook) |
ISBN 9781032108636 (hbk) | ISBN 9781032108698 (pbk) |
ISBN 9781003217473 (ebk)
Subjects: LCSH: Earthquake resistant design. | Failure time data analysis.
Classification: LCC TA658.44 .S3583 2022 (print) |
LCC TA658.44 (ebook) | DDC 624.1/762--dc23
LC record available at https://lccn.loc.gov/2021020844
LC ebook record available at https://lccn.loc.gov/2021020845

ISBN: 978-1-032-10863-6 (hbk)
ISBN: 978-1-032-10869-8 (pbk)
ISBN: 978-1-003-21747-3 (ebk)

DOI: 10.1201/9781003217473

Typeset in Times
by Newgen Publishing UK

Contents

Foreword

I am delighted to write the Foreword for the second edition of this book, which captures the recent advances on fundamentals of the "Endurance Time" methodology and its application to earthquake-resisting design of structures. Protection of buildings against earthquakes is of critical importance for seismic regions. In the conventional aseismic design, the structure is strengthened to sustain the expected earthquake ground shaking without failure and/or major damages. However, observations of structural failures during earthquakes revealed that most destruction occurs as a result of the gradual increase of damages and weakening of the building due to the earthquake-induced loads. In this book, a new innovative method for performance analysis of buildings to earthquake strong motion is presented. In this approach, the structure is subjected to a seismic excitation with an intensity that increases with time. The building damages are monitored until the collapse of the structure, and the corresponding Endurance Time is evaluated. The Endurance Time method provides a new realistic approach for the aseismic design of complex structural systems. The second edition revised, updated, and enhanced the materials in the first edition.

In my view, the authors of this book are the pioneering experts in this area and have been the developer of the Endurance Time methodology from its inception to formalizing its applications to the earthquake-resisting design of structures. I believe this book provides the readers with an in-depth understanding of the Endurance Time methodology for the aseismic design of buildings. The challenge that has been presented to the authors has been to introduce the fundamental aspect of the Endurance Time concept in a single textbook and bring it all the way to the actual design of buildings. The authors can be proud that they have managed to achieve such a task seamlessly. The description of the fundamentals of Endurance Time methodology is broken down to easy-to-understand exposition, beginning from the first principals and building up towards a thorough description of the method targeted for application to seismic response analysis of structures and evaluation of the associated damages. More importantly, the authors have provided useful best-practice solution methods that are accompanied by a practical approach in aseismic building designs. Whether readers view this text from either the fundamentals of Endurance Time methodology or the earthquake engineering side, the book provides readers with a clear understanding of the comprehensive treatment of the structural responses with the innovative approach of Endurance Time methodology for designing buildings against earthquakes. This book is certainly an enthusiastic celebration of three basic elements – theory, modeling, and practice – in handling a multitude of structural design issues under earthquake excitations.

Goodarz Ahmadi
Clarkson University
Potsdam, New York

Preface

Endurance Time Method (ETM) is a relatively new approach to seismic assessment of structures. This method has been developed into a versatile tool in the field of earthquake engineering, and its practical applications are expected to rapidly expand in the near future. ETM is a dynamic analysis procedure in which intensifying dynamic excitations are used as the loading function. ETM provides many unique benefits of its own among available seismic analysis procedures. First of all, it is a response history-based procedure. Thus, ETM inherits the capabilities of true dynamic procedures in analysis of systems with complexities. These include issues arising from nonlinear behavior and irregularity in the system that defies simplification beyond certain limits. Second, the ET method considerably reduces the computational effort needed in typical response history analyses. This reduction is roughly in the order of 10- to 100-fold depending on the number of ground motions and intensity levels intended in equivalent time history analysis. A huge computational effort is still a factor in seismic analysis of large and/or complex dynamic systems that makes conventional response history analysis of acceptably detailed models impractical. The third important advantage of the method lies in the simplicity of its concept. Conceptual simplicity makes it a great tool for preliminary response history analysis of almost any dynamic structural system. By applying an intensifying dynamic loading, many of the modeling issues can be readily identified and corrected. The structural analyst is mostly relieved of the complexities involved in ground motion selection and scaling issues.

Some of the most important areas of application of ETM are in the fields of seismic design optimization, value-based seismic design, and experimental studies. Seismic design optimization usually requires a huge amount of computational effort, and ETM can pave the way toward practical use of response history-based analysis improving the reliability of the design. Design optimization based on lifecycle cost evaluation and, more generally, value-based design optimization also involves too many repetitive response history analyses, and ETM can be useful in this regard. Another area of application of the ET method – an area that has been less explored until now – is in experimental investigations in shaking-table labs. While the number of analyses may be a less critical factor in computation, it is indeed among the most critical factors in experimental studies, and ETM can be very useful in these studies.

This book is aimed to serve as a coherent source of information for students, engineers, and researchers who want to familiarize themselves with the concepts and put the concepts into practice. The chapters of the book are organized in a relatively independent manner. After reading Chapter 1, the reader can essentially continue with any chapter that covers his or her topic of interest. However, readers who want to develop a deep understanding of the method and its development are encouraged to cover all chapters in order.

We are indebted to Professor G. Ahmadi of the University of Clarkson for reading the manuscript and offering his scientific feedback. We would like to thank Professor Mohammad Noori of the California Polytechnic State University for ongoing encouragement. Also, we would like to express our gratitude to Joseph Clements and Lisa Wilford from Taylor & Francis Group, for their professionalism and support in the process of publication.

This work could not be materialized without the contribution of our graduate students at Sharif University of Technology. The authors would like to specially thank Dr. H.T. Riahi, Dr. M.C. Basim, Dr. V. Valamanesh, Dr. A. Nozari, Dr. A. Mirzai, Dr. A. Bazmuneh, Dr. M. Mashayekhi, and Dr. M. Foyuzat whose excellent contributions in the development of ET method appear in this book. The authors would also like to thank the staff of the *Scientia Iranica* journal for their indispensable efforts and cooperation during preparation of this work.

— H.E. Estekanchi and H.A. Vafai

Biographies

Homayoon Estekanchi is a professor of civil engineering at Sharif University of Technology. He received his PhD in civil engineering from SUT in 1997 and has been a faculty member at SUT since then. He is a member of the Iranian Construction Engineers Organization, ASCE, the Iranian Inventors Association, and several other professional associations. His research interests include a broad area of topics in structural and earthquake engineering with a special focus on the development of the Endurance Time Method and Value-based Seismic Design.

Hassan Vafai has held a professorship in civil engineering at different universities, including Sharif University of Technology, Washington State University, and the University of Arizona. He was founder and editor-in-chief of *Scientia*, a peer-reviewed international journal of science and technology. Throughout his professional career, he has received numerous awards and distinctions, including emeritus distinguished engineer by the National Academy of Sciences, Iran; an honorary doctorate by the Senatus Academicus of Moscow Region State Institution of Higher Education, and the Order of Palm Academiques, awarded by the French Ministry of Education, Research and Technology.

1 Introduction to the Endurance Time Method

1.1 INTRODUCTION

The basic objective of seismic design is to provide the structure with an appropriate safety margin against failure when subjected to strong earthquakes (Estekanchi 1994).[1] The common philosophy of most well-known seismic design codes is to achieve the dual goal of keeping the nonstructural damage to a minimum in the case of service-level earthquakes and, also, to prevent structural failure in the case of collapse-level earthquakes (Moghaddam 2002, Newmark 1971). Early observations of structural failure during earthquakes revealed that most could be attributed to the weakness of structures in sustaining imposed lateral loads and displacements. This observation formed the premise of well-known earthquake design criteria based on lateral load, also known as the static seismic design method. In the static lateral load method, the structure is designed to resist a minimum lateral load specified by the code. In this way, minimum lateral strength and stiffness are provided and lateral displacements are limited (Moghaddam 2002; Estekanchi, Vafai, and Shahbodaghkhan 2003).

Extensive research in the field of earthquake engineering has revealed many deficiencies and shortcomings in the traditional method of static seismic analysis (Chandler and Lam 2001). According to the static method of analysis, structures with higher lateral strength and stiffness are superior to their less-stiff and less-strong counterparts. However, experimental and analytical investigation shows that this is not always the case (Shortreed, Seible, and Benzoni 2002; Bertero and Bertero 2002; Moehle and Elwood 2003). In fact, there are cases in which reducing lateral stiffness results in better seismic performance. The concept of seismic base isolation is one example (Moghaddam and Estekanchi 1999; Estekanchi 1993).

These apparent shortcomings in traditional seismic design, along with remarkable developments in the field of information technology and the availability of vastly improved analytical tools, have led researchers and engineers to develop more rational and consistent methods for earthquake engineering (Chopra and Geoel 1999; Chopra and Goel 2003; Medhekar and Kennedy 2000; Gupta and Kunnath 2000). In this respect, Performance-Based Seismic Engineering (PBSE) has gained increased interest among practitioners in the earthquake engineering field (Kelly and Chambers 2000; Judi, Davidson, and Fenwick 2003). Development of these new methods and criteria should be mainly attributed to the amazing improvement in computational tools that have made possible the solution of sophisticated nonlinear models (Vafai, Estekanchi, and Ghadimi 2000). The use of static push over analysis has become a standard practice in structural engineering design offices, and the application of nonlinear time history analysis is also gaining popularity quite rapidly. Thanks to these recent developments, it has become possible for the structural analyst to incorporate the most significant nonlinear material and geometric behavior into the model and, thus, perform a more realistic analysis of structural behavior during earthquakes.

In recent years the development of new analysis tools has provided the infrastructure for the development of new methods in structural engineering. Endurance Time (ET) method is for seismic analysis and evaluation of structures (Estekanchi, Vafai, and Sadeghazar 2004; Estekanchi et al. 2020). In this chapter, the concepts of the Endurance Time method and prospective methods to implement it are discussed. This method is first evaluated by applying the concept to linear single- and multi-degree of freedom (SDOF and MDOF) systems. Extension into nonlinear and more complicated applications will be discussed in the next chapters.

1.2   THE ENDURANCE TIME CONCEPT

The concept of the ET method can be best explained by considering a hypothetical shaking table experiment. Assume that three model buildings with unknown seismic resistance characteristics are to be investigated with reference to their resistance against collapse in severe earthquakes. Imagine that these models are put on the shaking table and fixed to it. The experiment starts by subjecting the buildings to random vibration with gradually increasing intensity. In the beginning (e.g., at t = 5 sec) the amplitude of shaking is quite low, so all three buildings vibrate but remain stable, as shown in Figure 1.1a. As the amplitude of vibration is increased (say, at t = 10 sec), a point is reached when one of the buildings collapses. Assume that this happens to be Model Number 1, as shown in Figure 1.1b. As time passes and the vibration amplitude is further increased (say, at t = 20 sec), the second structure fails. Assume this to be Building Number 3, as shown in Figure 1.1c. Further, let us assume that Building Number 2 happens to be the last building to fail in this hypothetical experiment.

Now, based on this experiment, it can be concluded that Building Number 2, which endured longer, has the best performance against collapse when subjected to a dynamic load. On the other hand, Building Number 1, which failed the soonest, can be labeled as the worst performer when subjected to dynamic load. Note that the judgment is based on their endurance time, that is, the maximum time each building remained stable. This judgment is reached without any reference to the building strength or stiffness or other dynamic characteristics, due to the fact that the dynamic response is being directly observed. If the characteristics of the applied dynamic excitation can be correlated to ground motions, then such experiments seem to show a direct and relevant measure of seismic resistance. This is the basic concept of the ET method in evaluating the performance of buildings subjected to earthquakes.

In ET method, buildings are rated according to the time they can endure a standard calibrated intensifying accelerogram. Higher endurance is interpreted as better performance. Standard minimum performance requirements can be set to be used as

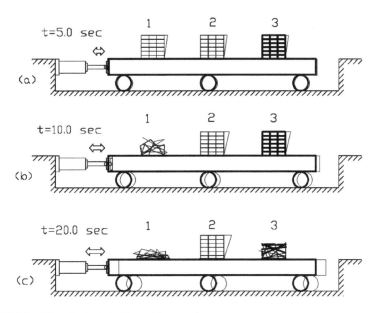

FIGURE 1.1 Hypothetical shaking table experiment.

the design criteria. The idea of the ET method is somewhat similar to the method used by cardiologists to evaluate the condition of the cardiovascular system, known as the stress test. In a stress test, the patient is asked to walk on a treadmill with a variable slope and speed. The test starts with low-slope and low-speed conditions. During the test, the slope and speed are increased gradually, while the physical and biological condition of the patient, such as blood pressure, heartbeat rate, and so on, are monitored. The test is continued until signs of distress or abnormal conditions are observed. The state of the patient's cardiovascular system is then judged on the basis of the maximum speed and slope level that could be tolerated.

In the ET method, nearly the same concept is applied. The idea behind the ET method is roughly to put the structure on a ramp-like accelerogram and see how far it can go. The structure is subjected to a calibrated accelerogram with intensifying dynamic demand. The specified performance indexes are monitored as the applied dynamic load follows an intensifying pattern. The performance of the structure is then judged on the basis of the time at which the damage limit states are exceeded. This concept has been explained graphically in Figure 1.2. Consider a damage index curve (e.g., maximum drift, plastic energy, etc.) for a typical structure subjected to an intensifying accelerogram, as shown in Figure 1.2.

If the limit value for the specified damage index is specified to be 1.00, then it can be concluded from Figure 1.2 that this structure has endured the accelerogram up to about the 12th second. Moreover, assume that the accelerogram has been calibrated and the design criterion is that it should endure dynamic intensity reached at the 10th second. As can be seen from this figure, the damage value is about 0.82 at t = 10.0 seconds, that is, below the value limit, thus, one can conclude that the structure has met

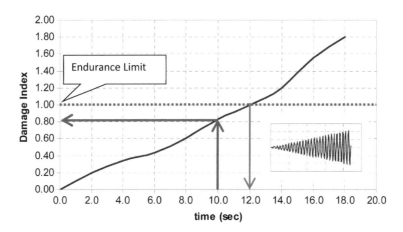

FIGURE 1.2 Damage curve against time for a typical structure subjected to intensifying accelerogram.

this design criteria. The analysis should not necessarily be limited to a single design criterion. Various criteria and different damage indexes and structural performances at different intensity levels can be considered simultaneously in order to reach more conclusive assessment of structural performance.

Now, to apply this idea in practice, one needs to propose a method for implementing its conceptual explanation. Direct physical testing of actual structures as described earlier, while theoretically possible, is not practical for engineering application considering time and cost implications. Therefore, numerical and analytical approaches are to be adopted. Analytical software capable of modeling and predicting structural behavior up to the complete collapse point are available, and will be discussed in subsequent chapters. In this chapter, a basic implementation of the ET method is presented by making some simplifying assumptions and by making use of simple analytical tools. Complete numerical implementation of the method, which ideally should follow the nonlinear response up to collapse level, will be left to Chapter 2 and will apply the latest generation of structural engineering software and computer hardware capable of performing sophisticated analysis.

Rudimentary implementation of the concept is based on three fundamental principles, that is, dynamic input, structural model, and endurance criteria. These will be discussed next.

1.3 ENDURANCE TIME EXCITATION FUNCTIONS

The choice of an appropriate dynamic input is fundamental to the successful implementation of the ET concept. The ideal input function is one which results in higher consistency and the best correlation between the ET analysis results and response history analysis results of structures subjected to earthquakes. Determination of the optimal dynamic excitation function is the subject of ongoing research work. In this chapter, simplistic Endurance Time Excitation Functions (ETEFs) produced based

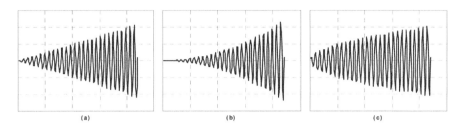

FIGURE 1.3   Input profile functions; (a) Linear, (b) Increasing rate, (c) Decreasing rate.

on engineering judgment and some elementary calculations will be proposed so that the concept and application of the ET method can be explained. These ETEFs will be in the form of ground acceleration functions, as commonly used in earthquake engineering.

An important issue in determining dynamic input is the profile of amplitude (or intensity) intensification. As shown in Figure 1.3, the amplitude increase profile can take various forms. After considering several possible alternatives, the authors came to the conclusion that a linear profile, as shown in Figure 1.3a, is most suitable for the purpose of initial development. In this profile, maximum acceleration is directly proportional to time. Determination of an optimum profile that results in better correlation and consistent results is an interesting subject to investigate separately.

Another important consideration in ET analysis is the specification of the dynamic input function itself. The most basic form of dynamic input is a simple harmonic. But this kind of input has obvious disadvantages because of poor frequency content consisting of only one harmonic. The structures that have natural vibration periods near the input frequency will experience high dynamic magnification and, thus, will be penalized too much. Considering this issue, a random vibration input with a frequency content resembling that of white noise has been used as the starting point for generating of intensifying accelerograms. As will be explained later, the frequency content is later modified to better correspond to what is expected in a real earthquake. The accelerograms generated by this procedure seem to be good enough for investigation of the ET concept and will be applied in this study. These are called the first generation of ETEFs. More advanced forms of ETEFs will be explained in next chapters. Producing optimal ETEFs for various applications in seismic analysis, design, and assessment of structures is an open problem to be researched in the future.

1.3.1   GENERATION OF INTENSIFYING ACCELEROGRAMS

The first-generation ETEFs to be used as dynamic input in the ET method is produced using random numbers with a Gaussian distribution of zero mean and a variance of unity (Clough and Penzien 1993). A stationary random accelerogram generated using $\Delta t = 0.01$ and $n = 2^{11} = 2048$ with PGA = 1 is shown in Figure 1.4. Duration of the accelerogram is equal to $\Delta t \times n = 20.48$ seconds.

The frequency content of the random accelerogram, that is statistically similar to white noise, is then modified in order to resemble actual earthquake accelerograms.

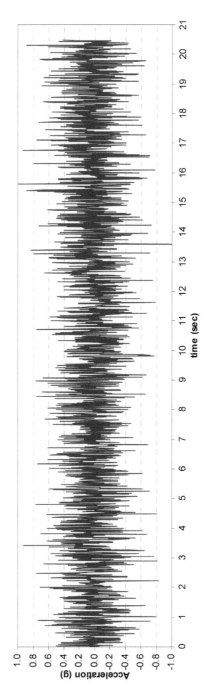

FIGURE 1.4   A typical random accelerogram with PGA = 1.

For this purpose, filter functions given by formulas (1.1) and (12) are applied to the random accelerograms. Application of these filter functions is explained by Clough and Penzien (1993).

$$H1(i\varpi) = \frac{1 + 2i\xi_1 \dfrac{\varpi}{\omega_1}}{\left(1 - \dfrac{\varpi^2}{\omega_1^2}\right) + 2i\xi_1 \dfrac{\varpi}{\omega_1}} \tag{1.1}$$

$$H2(i\varpi) = \frac{(\dfrac{\varpi}{\omega_2})^2}{\left(1 - \dfrac{\varpi^2}{\omega_2^2}\right) + 2i\xi_2 \dfrac{\varpi}{\omega_2}} \tag{1.2}$$

Where $\omega_1 = 2\pi/0.5$, $\xi_1 = 0.2$, $\omega_2 = 2\pi/0.1$, $\xi_2 = 0.2$ are used. It should also be noted that frequencies higher than 100 Hz are filtered out. A sample frequency content resulting from the accelerogram shown in Figure 1.4 is depicted in Figure 1.5.

The frequency content is then further modified so as to make the resulting response spectra compatible with typical seismic code Response Accelerograms. Codified design response spectra or ground-motions response spectra can be used for this purpose based on the desired application. In this chapter, the response spectra of the Iranian National Building Code (Standard 2800) (Estekanchi, Vafai, and Shahbodaghkhan 2003) has been used as a sample. The resulting frequency content of a response-spectra-compatible accelerogram is shown after several cycles of step-wise modification, in Figure 1.6.

The acceleration response of the complete accelerogram is depicted in Figure 1.7 in comparison with the sample code design spectrum. The convergence is assumed to be good enough for the purpose of explanation of the ET concept. More accurate calibration of the ETEFs will be explained in following chapters.

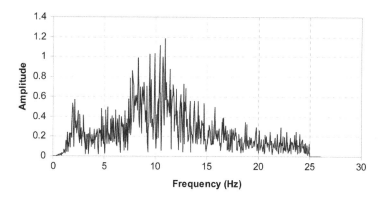

FIGURE 1.5   Filtered frequency content of the accelerogram.

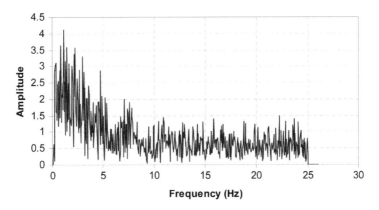

FIGURE 1.6 Modified frequency content of the accelerogram.

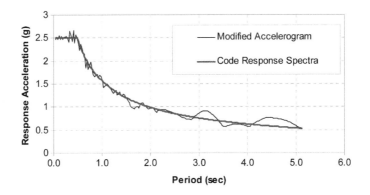

FIGURE 1.7 Response of modified accelerogram compared to codified value.

In the next step, the acceleration values are adjusted for target values of velocity and acceleration at record end – in this case arbitrarily set to zero. Finally, the acceleration values are multiplied by a profile function which, in this study, is a linear one starting from zero and reaching a value of 1.00 at t = 10 seconds. Three accelerograms have been generated using this procedure. These accelerograms will be referred to as acc1, acc2 and acc3. Accelerogram acc1, which is obtained from the original accelerogram in Figure 1.4, is shown in Figure 1.8. These belong to the first generation of ET excitation functions.

1.4 ANALYSIS METHOD

Collapse analysis of the structures is still considered a challenging task for structural engineers involved in the numerical analysis of structures. Few programs are capable of conducting such analysis with reasonable accuracy. Also, the experimental evidence to verify the results of such analyses is still quite limited. The minimum modeling requirement for collapse analysis is nonlinear material

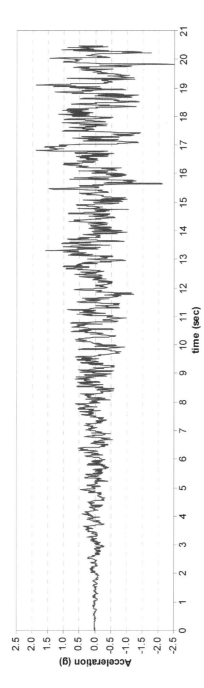

FIGURE 1.8   Intensifying accelerogram acc1.

behavior, including material degradation due to cyclic behavior, large strain, and fracture. Analysis should also include the effect of large deformations, buckling, and collision. Therefore, for the purpose of illustrating the ET method, the simplest procedure for dynamic analysis is considered in this chapter, that is, linear systems. The Newmark linear method has been applied for the analysis (Chopra 1995). It should be evident that, as far as the concept of the ET method is considered, the procedure remains almost the same, even for complex multi-degree-of-freedom models involving nonlinear features.

1.5   ENDURANCE CRITERIA

The numerical definitions of structural collapse and failure do not yield to a straight-forward formulation. An intuitive failure criterion can be defined as the displacement of the center of mass of the structure to a lower level from its initial position. However, analytical implementation of such criteria is still considered to be very demanding in terms of computational effort and modeling complexities. The state of the art in this regard is to define the structural damage in terms of damage indexes. Various damage indexes have been defined and proposed by researchers in recent years. Seismic codes that are based on performance-based design usually propose a certain damage index and set maximum acceptable values for it. In this way, the structure is assumed to have collapsed when its damage index exceeds specified code limits. The same simplified method shall be used for the purpose of this study, as described in the next sections. Endurance Time is defined as the time it takes for the specified damage criteria to reach its limit value when the structure is subjected to intensifying accelerograms. For example, in the static method of earthquake-resistant design, structures are required to be designed according to a specified base shear, which is proportional to certain peak lateral ground acceleration. Also, limits to maximum lateral displacements and drift are usually specified. These can be considered as the simplest forms of damage indexes. In the discussion that follows, basic parameters such as "max response acceleration," "max displacement response," and "max interstory drift" have been used for explanation of the idea. It will be up to the structural designer/analyst to choose an applicable damage index that is more appropriate for the structure under investigation. It should be evident from the discussion that many different damage indexes can be considered simultaneously and at various intensity levels in the procedure. Practical implications of applying different damage indexes for ET analysis will be explained in later chapters.

1.6   APPLICATION OF INTENSIFYING EXCITATION TO LINEAR SDOF SYSTEMS

The most significant characteristics of a single-degree-of-freedom (SDOF) structure is its natural frequency of vibration. In this section three different linear SDOF systems with natural periods of vibration equal to 0.1, 0.5, and 1.0 seconds are studied. These structures will be hereafter referred to as ST01, ST05, and ST10. ST01 roughly pertains to a stiff single-story masonry building, while ST05 and ST10 have a period typical of three- and nine-story steel frame buildings, respectively. Damping ratio will be assumed to be 0.05 of the critical value, as commonly assumed in dynamic analysis.

Our analysis starts with application of the accelerograms produced in previous section to ST05. The acceleration response for ST05, subjected to the three generated accelerograms, is shown in Figure 1.9. As expected, peak acceleration increases with time as the input acceleration is intensified. In Figure 1.10, maximum acceleration as a function of time has been plotted along with average value and a linear fitting curve. As can be seen in this figure, in spite of the fact that the generated accelerograms are compatible with the same design response spectra and have a similar frequency content, the dynamic response can be different at specific time intervals. For example, consider that one wants to specify the time at which the structure has experienced a maximum acceleration of 1g. From Figure 1.10, it can be seen that for acc1, this occurs at t equals to about 11.5 seconds, while for acc2, this occurs at t = 16 seconds.

This could be expected considering the random characteristics of applied accelerograms. The effect of randomness can be reduced if the average value from multiple analyses is used as the best estimate for expected response at desired load intensity. Various combinations of averaging methods and analysis numbers can be used to achieve desired level of convergence. A curve-fitting procedure can also be applied along with averaging to reduce randomness effects. In this example, using the average from three ET analyses and a linear fitting curve, the time at a = 1g can be seen to be about 14 seconds.

The displacement response of ST05 subjected to different accelerograms has been depicted in Figure 1.11. As can be seen in this figure, the displacement response is also an increasing function of time, as expected. The pulsating characteristic of the displacement response should receive due attention. These pulsations can result in the maximum response to remain constant during a relatively long period of time, making it difficult to interpret the result of analysis regarding the time corresponding to a certain response level. This problem can be avoided by using several accelerograms, along with an appropriate averaging method. Even though, in this chapter, the number of accelerograms to be averaged has been set to three as a reasonable practical number, it should be clear that the desired accuracy and convergence can be achieved by considering any number of accelerograms.

Maximum and average accelerations are shown in Figure 1.12. Maximum displacement is considered to be a simple and effective damage-estimation criterion. In multistory buildings, maximum displacement is usually proportional to maximum drift, which is another significant response characteristic that can be related to building damage. Consider that the maximum tolerable displacement for this building has been set to 4cm. Based on Figure 1.12, one can conclude that the building can endure the prescribed accelerogram up to t = 11 seconds.

A summary of the results for the maximum acceleration of ST01, ST05, and ST10 is given in Figure 1.13. In general, the magnitude of acceleration experienced by all three structures in this example turns out to be nearly the same, with that of ST05 being higher for almost the entire time range, and that of ST10 being lower in the time interval from 7.5 to 12.5 seconds. These could be expected considering the shape of the codified response spectrum and the frequency content being amplified near to T0 = 0.5, according to the assumed soil conditions.

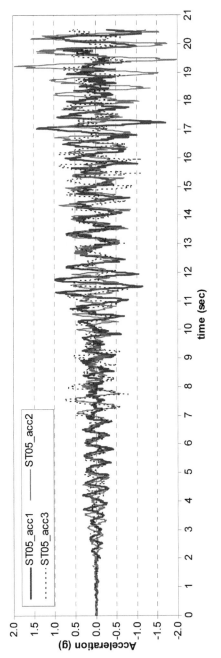

FIGURE 1.9   Acceleration time history for ST05.

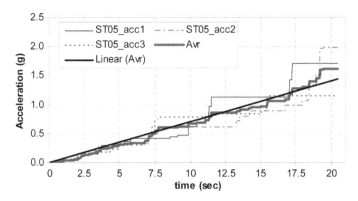

FIGURE 1.10 Maximum acceleration and average for ST05.

Displacement responses of these systems have been depicted in Figure 1.14. As can be seen in this figure, there is a marked distinction between the displacement responses of sample structures. Significant insight on the structural performance of these structures can be gained through studying such curves as of those in Figure 1.14. For example, consider that the maximum tolerable displacement for all these structures was to be limited to 4cm. It can be concluded from Figure 1.14 that the ET for ST10 is about 6 seconds, for ST05 about 11 seconds and for ST01 more than 20 seconds. Now, by specifying the required endurance time of 10 seconds, it can be concluded that ST01 and ST05 are acceptable – that is, can endure the specified dynamic demand, while ST10 is rejected.

As another example, consider that ST05 and ST10 are three- and nine-story buildings with heights of 10m and 30m, respectively, and the failure criteria is set to be a maximum drift of 0.005. If one roughly assumes the vibration mode to be linear, maximum displacements will be $0.005 \times 10 = 0.05$m and $.005 \times 30 = 0.15$m, respectively. Now, by referring to Figure 1.14, it can be seen that ST05 reaches the value limit at about t = 15 seconds, while ST10 reaches its value limit at t = 15.5 seconds. Thus, the ET for both structures is nearly the same and, if one sets the required ET to 10 seconds, then both structures will be considered acceptable. These discussions are of course only for explanatory purposes and more realistic and in dept discussions will be presented in the following chapters.

In the earlier discussion, the linear analysis of a single-degree-of-freedom system is considered for the purpose of describing the basic idea behind the ET method. It should be clear from this discussion that the analysis can be readily extended to non-linear and multi-degree-of-freedom systems as well. The essence of the ET method lies in the definition of standard intensifying accelerograms and appropriate damage criteria. It is interesting to note that the concept of ET can also be readily applied in experimental dynamic investigation of structures. In this case, the most realistic performance criteria, that is, actual failure of the structure, can be considered the endurance limit as measured against time. Another advantage of the ET method, as compared to other dynamic analysis methods, is in its applicability to experimental shaking table studies. In these cases, the cost and required resources usually eliminate

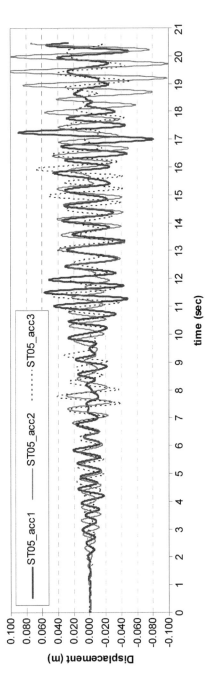

FIGURE 1.11 Displacement time history for ST05.

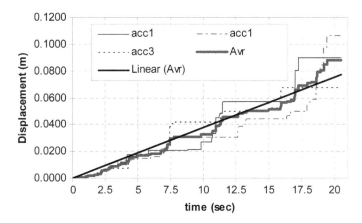

FIGURE 1.12 Maximum displacement and average for ST05.

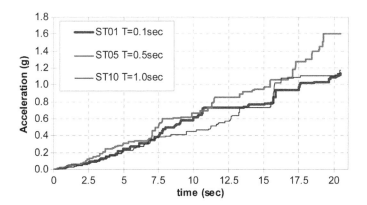

FIGURE 1.13 Maximum average accelerations.

the possibility of conducting the experiment several times with different levels of excitation. Using the concept of ET, a single-run experiment can be used to obtain the desired results.

1.7 APPLICATION TO MDOF SYSTEMS

In this section, the application of the ET method in dynamic analysis of MDOF systems is explained by considering a three-story steel moment frame example. A three-story steel frame, named as "f3" in Figure 1.15, has been designed in accordance with conventional loading and analysis procedures appropriate for ordinary moment frames in high seismic activity zones. For comparison, another frame with similar properties, but utilizing steel sections with lower stiffness and strength, has also been designed. This weaker frame is named f3w in Figure 1.15. We intend to analyze these frames using the ET concept and study the results. For multistory frames, interstory drift

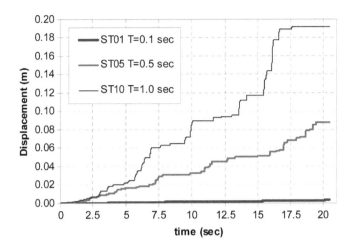

FIGURE 1.14   Maximum average displacements.

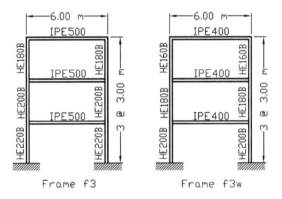

FIGURE 1.15   Three-story steel moment frames.

has proved to be a very significant and convenient measure of building damage and performance criteria. We shall use this criterion for the purpose of explanation of the method. It should be clear that other criteria based on strain energy, nonlinear hysteresis behavior, and so on. can also be applied with few further complications. The only limitation in the application of the ET method lies on the computational capability of the analysis program being used.

Frame f3 has been subjected to accelerogram acc1, and the time history of story drifts has been depicted in Figure 1.16. It can be seen from this figure that the story drifts are nearly the same for all stories in most of the time history and, also, they reach peak values at almost the same times. This is because in MDOF frame structures, Mode Number 1 is usually the predominating deflection mode. Higher vibration modes have had much less pronounced effect in peak values for interstory drifts in this case.

Frame f3 also has been subjected to accelerograms acc2 and acc3, and the maximum absolute values of drifts along with their average value are depicted in

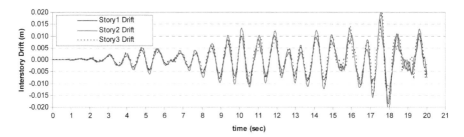

FIGURE 1.16 Story drifts time history for frame f3 subjected to acc1.

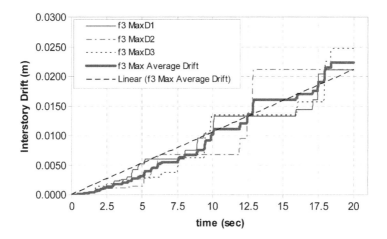

FIGURE 1.17 Maximum story drifts for frame f3 subjected to acc1-3 accelerograms.

Figure 1.17. As expected, the maximum drift increases with time and, by considering a limiting value for drift, the endurance time can be evaluated. Due to the randomness of input and resulting response, the peak value can remain constant for a prolonged period of time. For example, maximum drift has remained constant from $t = 10$ to $t = 16$ seconds for f3 subjected to acc1, as can be seen in Figures 1.16 and 1.17. Considering the irregular fluctuation in the drift values for the frame subjected to each accelerogram, the average value can be used to reduce the inaccuracy that can result from the random nature of response.

A curve fitting method can also be used in order to better define the ET value. In Figure 1.17, a linear trend line has been added for this purpose. In linear systems subjected to linearly intensifying accelerograms, we know that the response is roughly linear. This justifies the use of a linear trend line as the fitting curve. However, it should be clear that, when applying the ET method to non-linear problems, a more appropriate fitting method, such as spline or polynomial curves, should be used.

Now, considering a limiting value of 0.005 for interstory drift, as a typically recommended maximum value, the limiting drift for frame f3 with a story height

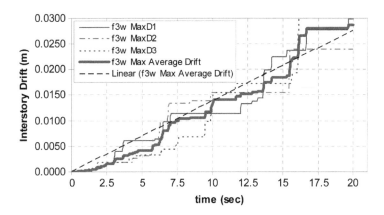

FIGURE 1.18   Maximum story drifts for frame f3w subjected to acc1-3 accelerograms.

of 3.0m will be 0.005 × 3.0 = 0.015m. Thus, we can conclude from Figure 1.17 that frame f3 has endured our acc1-3 set of accelerograms up to time t = 13.5 seconds. If we have standardized these accelerograms for the specified site and had specified a minimum ET of say 10 seconds, we would then conclude that frame f3 passes our seismic design criteria. Generating a standard set of intensifying accelerograms to be used as design criteria in the ET method will be explained in the next sections.

The result of the same analysis as applied to frame f3w – that is, the weak version of f3 – is summarized in Figure 1.18. The general trend for f3w is the same as f3. It should be noted that the resulting drift values are generally higher for f3w. This could be expected considering the lower stiffness of frame f3w. Considering the limiting value of 0.005 for interstory drifts as earlier, it can be concluded that the ET for f3w is about 10.5 seconds.

Analysis of results for frames f3 and f3w are compared in Figure 1.19. The significant point in this figure is the ET for f3w being clearly lower than that of f3. On the other hand, the weakness of f3w as compared to f3 shows up in its response to the intensifying accelerograms. Again, if we assume that the accelerograms have been somehow standardized, and a minimum ET of 10.0 seconds has been specified to the building site, it can be concluded that f3w passes our seismic design criteria. If, however, the minimum ET was specified to be 12.0 seconds, f3 would pass the design criteria, while f3w would fail.

In case of the simple example considered earlier, one could already predict the analysis result by considering that f3w is less stiff than f3. However, in case of more complex models, particularly the models involving several sources of nonlinearity, it may of course not be possible to predict the ET and the advantage of one design over the other. For example, consider that to be studied is the effectiveness of various energy-absorbing devices in reducing the maximum drift of a structure. These types of analyzes are usually encountered when considering seismic retrofit studies. It should be clear from the earlier discussion that by applying the ET method, a conclusive dynamic analysis result can be readily achieved.

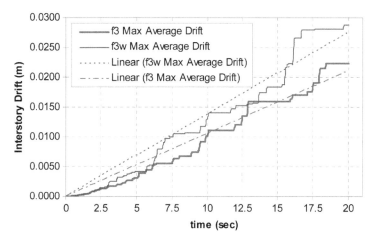

FIGURE 1.19   Maximum average story drifts for frames f3 and f3w.

1.8   SUMMARY AND CONCLUSIONS

In this chapter, a new approach to performance-based earthquake analysis and design has been introduced. In this method, the structure is subjected to dynamic excitations that impose increasing dynamic demands that increase with time. Damage indexes of interest are monitored through time as the dynamic load intensity is increased. The elapsed time until collapse-level damage or other performance limits are reached is called the endurance time of the structure. ET is defined as the length of the time interval from start to failure point (or limit point), that is, the time at which the damage index reaches its maximum tolerable value. Longer ET implies better or improved performance. The notion of endurance time is rather arbitrary in this kind of analysis. Standard accelerograms can be calibrated and a minimum endurance time can be specified at desired intensity levels to be used as the design or decision criteria. In this way, the structure is considered acceptable or unacceptable by its ET being higher or lower than the benchmark value(s).

The ET method provides a unified and objective approach to seismic analysis and design of structures, a method that can be applied in numerical and experimental investigations regardless of model complexity. Also, the method is readily applicable to complex nonlinear systems. The definition and calibration of standard intensifying accelerograms is a key issue in this method. A simple approach to generating required accelerograms has been discussed. Three such accelerograms were applied in the simple case of linear elastic single-degree-of-freedom systems. Methods for generating improved accelerograms to achieve better convergence and practical implications in cases of nonlinear MDOF systems are presented in the next chapters.

NOTE

1 *Chapter Source*: Estekanchi, H.E., A. Vafai, and M. Sadeghazar. 2004. "Endurance Time Method for Seismic Analysis and Design of Structures," *Scientia Iranica* 11, no. 4, pp. 361–370.

REFERENCES

Bertero, R.D., and V.V. Bertero. 2002. "Performance-Based Seismic Engineering: The Need for a Reliable Conceptual Comprehensive Approach." *Earthquake Engineering & Structural Dynamics* 31, no. 3, pp. 627–652.

Chandler, A.M., and N.T.K. Lam. 2001. "Performance Based Design in Earthquake Engineering: A Multidisciplinary Review." *Engineering Structures* 23, no. 12, pp. 1525–1543.

Chopra, A.K. 1995. "Dynamics of Structures: Theory and Applications to Earthquake Engineering." *Prentice Hall International Series in Civil Engineering and Engineering Mechanics*. Englewood Cliffs, NJ: Prentice Hall.

Chopra A.K., and R.K. Geoel. 1999. "Capacity Demand-Diagram Methods for Estimating Seismic Deformation of Inelastic Structures: SDF Systems." Report No. PEER-1999/02, Pacific Earthquake Engineering Research Center, College of Engineering, University of California, Berkeley.

Chopra A.K., and R.K. Goel. 2003. "A Modal Pushover Analysis Procedure to Estimate Seismic Demands for Buildings: Summary and Evaluation." *Fifth National Conference on Earthquake Engineering*, 26–30. Istanbul.

Clough, R.W., and J. Penzien. 1993. *Dynamics of Structures*. McGraw-Hill.

Estekanchi, H.E. 1993. "A Review of History, Performance and Application of Seismic Isolators (in Farsi)." *Omran Magazine of Sharif* 10, pp. 42–44.

Estekanchi, H.E. 1994. "Basics of Earthquake Engineering (in Farsi)." *Omran Magazine of Sharif* 15, pp. 26–29.

Estekanchi, H.E., M. Mashayekhi, H. Vafai, G. Ahmadi, S.A. Mirfarhadi, and M. Harati. 2020, October. "A State-of-knowledge Review on the Endurance Time Method." *Structures* 27, pp. 2288–2299.

Estekanchi, H., A. Vafai, and M. Sadeghazar. 2004. "Endurance Time Method for Seismic Analysis and Design of Structures." *Scientia Iranica* 11, no. 4, pp. 361–370.

Estekanchi, H.E., A. Vafai, and P. Shahbodaghkhan. 2003. "Comparative Evaluation of Braced Steel Frames Designed According to the First and Second Edition of the Iranian Code for Seismic Resistant Design of Buildings (in Farsi)." *Building Engineering and Housing Science* 1, no. 1, pp. 11–16.

Gupta, B., and S.K. Kunnath. 2000. "Adaptive Spectra-Based Pushover Procedure for Seismic Evaluation of Structures." *Earthquake Spectra* 16, no. 2, pp. 367–391.

Judi, H.J., B.J. Davidson, and R.C. Fenwick. 2003. "Displacement Focused Seismic Design Methods: A Comparative Study." *Proceedings of the 2003 Pacific Conference on Earthquake Engineering*, New Zealand Society for Earthquake Engineering, 8 pages, Paper No. 108.

Kelly, T.E., and J.D. Chambers. 2000. "Analysis Procedures for Performance Based Design." *12th World Conference on Earthquake Engineering*, New Zealand Society for Earthquake Engineering, Paper No. 2400.

Medhekar, M.S., and D.J.L. Kennedy. 2000. "Displacement-Based Seismic Design of Buildings: Application." *Engineering Structures* 22, no. 3, pp. 210–221.

Moehle, J.P., and K.J. Elwood. 2003. "Collapse Performance Prediction for RC Frame Structures." *Proceedings of the 2003 Pacific Conference on Earthquake Engineering* [electronic resource], New Zealand Society for Earthquake Engineering, 8 pages, Paper No. 154.

Moghaddam, H. 2002. *Earthquake Engineering: Theory and Application*. Teheran: Farahang Publications.

Moghaddam, H.A., and H.E. Estekanchi. 1999. "A Study of Off-Center Bracing Systems." *Journal of Constructional Steel Research* 51, no. 2, pp. 177–196.

Newmark, N.M. 1971. *Fundamentals of Earthquake Engineering.* Englewood Cliffs, NJ: Prentice-Hall.

Shortreed, J.S., F. Seible, and G. Benzoni. 2002. "Simulation Issues with a Real-time, Full-scale Seismic Testing System." Controversial Issues in Earthquake Engineering, Proceedings of First ROSE Seminar. *Journal of Earthquake Engineering* 6, Special Issue 1, pp. 185–201.

Vafai, A., H.E. Estekanchi, and G. Ghadimi. 2000. "The Role of Information Technology in Construction Industry (in Farsi)." *Iranian Journal of Engineering Education* 1, no. 4, pp. 23–37.

2 Properties of the Endurance Time Excitation Functions

2.1 INTRODUCTION

In the ET method, structures are subjected to specially designed intensifying accelerograms called "Endurance Time Excitation Functions" (or ETEF for short), and their seismic performance is judged based on the maximum time(s) up to which they can satisfy the desired endurance criteria (Estekanchi, Vafai, and Sadeghazar 2004). [1] The criteria in measuring ET can be selected, based on the problem, to be the value of basic design parameters such as maximum drift or displacement, maximum stress ratio, or any other desired parameter or damage. Since the excitation imposed on the structure is an increasing function with time, the maximum value of displacements, internal forces, and other response parameters also increase with time in ET analysis. In this chapter, some basic properties of ETEFs that can be interesting from the seismic assessment viewpoint will be studied. The observation of damage in buildings after severe earthquakes shows strong interdependency between some ground-motion parameters and the structural response. Because of the complexity of earthquake ground motions, identification of a single parameter that accurately describes all important ground motion characteristics is not possible.

It is found that Spectral acceleration (S_a) and Spectral absolute seismic input energy have the strongest correlation with the overall structural-damage indices. On the other hand, the PGA (Peak Ground Acceleration), CP (Central Period, defined as the reciprocal value of the number of positive zero-crossing per time unit of the seismic acceleration), and SMD (Strong Motion Duration) exhibit relatively poor correlation with the overall structural damage indices (Elenas 2000).

This chapter provides a review of the major characteristics of a set of ETEFs belonging to the second generation of ETEFs. ETEFs can be classified into various generations based on the concepts and procedures used in their generation (Estekanchi et al. 2020). The term "first generation" refers to those ET acceleration functions that were generated by using a heuristic approach and applying a linearly increasing profile curve directly to a filtered acceleration function without direct control over response parameters. The term "second generation" refers to those acceleration functions in which optimization procedures have been applied in order to produce linear spectrum compliant acceleration functions. These acceleration functions make use of either codified design spectra or ground-motion spectra as a template spectrum (Estekanchi,

DOI: 10.1201/9781003217473-2

Valamanesh, and Vafai 2007; Estekanchi, Arjomandi, and Vafai 2008). The results presented in this chapter are aimed at providing a better understanding of various characteristics of acceleration functions. This study is based on the first set of ETEFs in the second generation with a code name of ETA20a series. This series of ETEFs is produced by using design acceleration spectra of soil type II in standard 2800 of INBC (BHRC 2005). As will be shown, the spectral acceleration of the code spectra used as template for these ETEF series is significantly biased towards higher values as compared to real ground motions in the long-period range. This issue, besides the fact that these ETEFs are fitted to the code spectra only up to the 5 seconds, limits their applicability only to linear analysis domain (Estekanchi, Valamanesh, and Vafai 2007). Some properties of these series that are indicative of the series limits will be shown in the next section as well.

2.2   SECOND GENERATION OF ET ACCELERATION FUNCTIONS

In the first generation of ET acceleration functions presented in chapter 1, the process of generating them started from a random vibration accelerogram, similar to white noise that was modified by a filter in the frequency domain and then made compliant with a typical code design-response spectrum. The resulting stationary accelerogram was then modified by applying a linear profile function that made it intensify, with respect to peak accelerations, at different time intervals. These accelerograms served well for the purpose of demonstrating the concept of ET analysis, but could not be expected to result in quantitatively significant results (Estekanchi, Vafai, and Sadeghazar 2004).

In the present chapter, properties of the second generation of ET accelerograms are explained. In this generation, in order for the ET acceleration functions to somehow correspond to average code-compliant design-level earthquakes, the concept of the response spectrum has been more directly involved. As will be explained later, these ET acceleration functions are designed in such a way to produce dynamic responses equal to the code's design spectrum at a predefined time, t_{Target}, and therefore, it is possible to compare the performance of various structures with different periods of free vibrations using these ETEFs. A time plot of a typical ET acceleration function, produced by mentioned procedure, has been depicted in Figure 2.1.

To calculate the response spectrum of the ET acceleration function at each time – for example, t_1 – the ET acceleration function can be cut at t_1 and its response spectrum can be sketched versus the period of vibration. By this approach, the average of response spectra of three ET acceleration functions at $t = 5$ sec, $t = 10$ sec, and $t = 15$ sec are depicted in Figure 2.2(a). The target time for this set of ET acceleration functions has been set to 10 sec. This means that the response at $t = 10$ sec should match the codified value with a scale factor of 1.0. Also, if a linear intensification scheme is used, at $t = 5$ sec and $t = 15$ sec, response spectra of these acceleration functions should match 0.5 (i.e., 5/10) and 1.5 (i.e., 15/10) times of the standard codified values, respectively. As can be seen in Figure 2.2, the ETEF generating process used for these records is quite successful in converging to the target values. These procedures will be discussed in chapter 4.

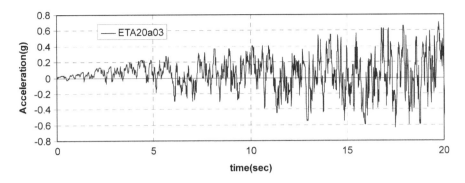

FIGURE 2.1   ETA20a03 acceleration function.

Displacement responses of these acceleration functions at various times are depicted in Figure 2.2(b). As expected, displacements follow the target values with almost the same level of dispersion as Spectral acceleration.

2.3 COMPARISON OF ET RESPONSE SPECTRUM WITH A REAL EARTHQUAKE

Even though ET acceleration functions are fundamentally different from earthquake records, it still helps to compare the level of various excitation parameters at different times with some real earthquake records set as some sort of reference values. The acceleration response spectrum is one of the most significant parameters from the structural engineering viewpoint.

As explained earlier, the template response spectra used in generation of the ETA20a series of accelerograms is that of the Iranian National Building Code (INBC) for stiff soil (type II). To compare this response spectrum with ground motions, the response spectra of seven earthquakes recorded on soil type C according to NEHRP provisions (FEMA 356 2000), listed in Table 2.1, are sketched in Figure 2.3. It should be noted that the characteristics of site class C of NEHRP are very similar to soil type II of the INBC standard 2800.

It is evident that at short periods the ET response spectrum conforms to INBC code as well as the average response spectrum of ground motions. However, at longer periods (T>0.4 sec) ET and INBC response spectra are considerably greater than the average of reference ground motions spectra. This is result of the safety factors used in the codified design spectrum in the long-period range and should be considered when comparing the characteristics of ground motions and code compliant accelerograms.

2.4 BASIC GROUND-MOTION PROPERTIES

A quick look at virtual ground velocity and displacement produced by ET records, as depicted in Figure 2.4, reveals some essential differences. As can be seen in these figures, equivalent displacements and velocities produced by ET records become too high after about 4 to 5 seconds. It should be noted that while a ground displacement

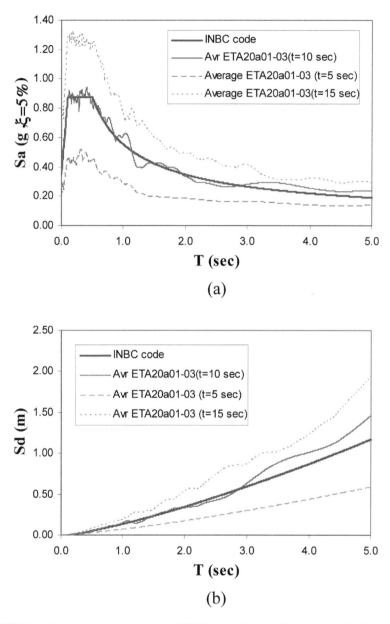

FIGURE 2.2   Average response spectra of ETA20a acceleration functions for ξ = 5 percent at different time. (a) Pseudo-acceleration, (b) Displacement Response.

of about 90m for ETA20a03 in Figure 2.4(b) cannot be compared to any real ground motion, it actually complies with the concepts of ET analysis and does not result in discrepancies for linear structures with a period of free vibration of up to about 5 seconds, which can be considered a quite long period for most building structures.

TABLE 2.1
Actual Record Events on Soil Condition C

Earthquake name	Date	Magnitude (Ms)	Station name	Station number	Component (deg)
Landers	06/28/92	7.5	Yermo, Fire Station	12149	0
Loma Prieta	10/17/89	7.1	Saratoga, Aloha Ave.	58065	0
Loma Prieta	10/17/89	7.1	Gilroy, Gavilon College	47006	67
Loma Prieta	10/17/89	7.1	Santa Cruz	58135	360
Loma Prieta	10/17/89	7.1	Anderson Dam,	1652	270
Morgan Hill	4/24/84	6.1	Gilroy #6, San Ysidro	57383	90
Northridge	1/17/94	6.8	Castaic, Old Ridge Route	24278	360

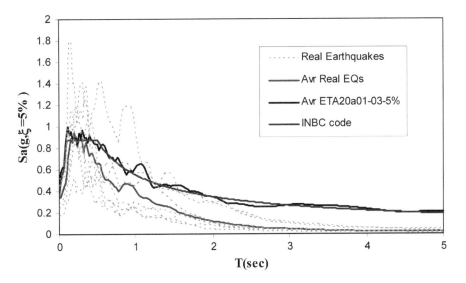

FIGURE 2.3 Comparison of ET response spectra (t = 10 sec), INBC code, and earthquakes response spectra.

As explained previously, the ETA20a series of acceleration functions have been optimized to fit with INBC code design spectra for stiff soil in the linear range. An evident conclusion from Figure 2.5 is that this set of ET accelerograms cannot be expected to yield reasonable results for structures with periods higher than 5 seconds. This conclusion also applies to structures with nonlinear behavior, where nonlinearity affects the structure in such a way as to elongate its effective period of vibration. Also, in the range of periods below 5 seconds, displacement demands that resulted from INBC design spectra are expected to be significantly higher than those from ground motions. Therefore, the ETA20a series of acceleration functions are not recommended for application in high period and nonlinear cases.

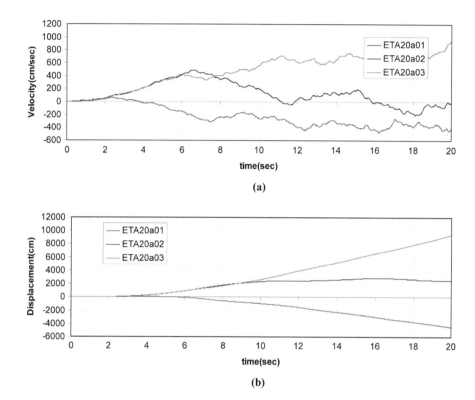

FIGURE 2.4 Velocity and displacement of ETA20a acceleration functions, (a) velocity, (b) displacement.

2.5   FREQUENCY CONTENT

Dynamic responses of structures are very sensitive to the frequency at which they are loaded. Earthquakes produce complicated loading with components of motion that span a broad range of frequencies. The frequency content describes how the amplitude of a ground motion is distributed among different frequencies.

The Fourier amplitude of ET acceleration functions up to 10 seconds is depicted in Figure 2.5. While the average Fourier amplitudes in ETA20a follow the general pattern of the ground motions, it shows significant differences at low frequencies, as expected.

Like response spectrum at frequencies between 2.5 and 10 Hz ($0.1<T<0.4$ sec), Fourier amplitude of ET acceleration functions is the same as average of ground motions. However, at long and short frequencies ($f<2.5$ Hz and $f>10$ Hz), Fourier amplitude of ET acceleration functions is greater than the average of seven earthquakes. An important note is that for almost all frequency ranges, Fourier amplitude from ET acceleration functions is greater than that of ground motions. It should be mentioned that the frequency content of ET acceleration functions is indirectly modified during the optimization process so that the response matches target values. As can be seen in Figure 2.5, frequency content is reasonable in the practical range of about 2.5 to 10

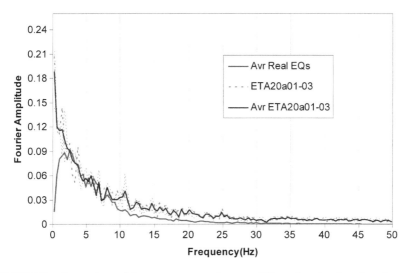

FIGURE 2.5 Comparison of frequency content between ET acceleration functions (at t = 10 sec) and ground motions.

Hz where numerical optimization has been carried out. Outside this range, the discrepancy is high. Therefore, for situations where effective frequencies can be outside this range, appropriate ET acceleration functions covering relevant frequency range should be used.

2.6 POWER SPECTRAL DENSITY

ET acceleration functions are inherently nonstationary, and their amplitude increases linearly with the time, therefore nonstationary approach should be used to describe the Power Spectral Density (PSD) of ET acceleration functions. Total intensity of ground motion with duration T_d is calculated in the time domain by Equation (2.1).

$$I_0 = \int_0^{T_d} [a(t)]^2 \, dt \tag{2.1}$$

From Parseval's theorem, I_0 can also be expressed in frequency domain as

$$I_0 = \frac{1}{\pi} \int_0^{\omega_N} [c_n]^2 \, d\omega \tag{2.2}$$

where $\omega_N = \pi / \Delta t$ is the Nyquist frequency and c_n is Fourier amplitude at frequency ω_N.

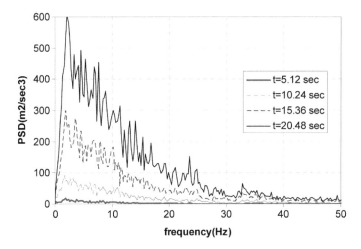

FIGURE 2.6 Power Spectral Density function for average of three ET acceleration functions ETA20a01-03.

PSD is defined such that

$$G(\omega) = \frac{1}{\pi T_d} c_n^2 \qquad (2.3)$$

The close relationship between the PSD function and the Fourier amplitude is apparent from the equation. The PSD is normalized by dividing its values by the area beneath it. Results from average of ET acceleration functions (Avr ETA20a01-3) and seven ground motions are depicted in Figures 2.6 and 2.7. It is obvious that spectral density increases parabolically with the time.

It is obvious from Figure 2.7 that ET acceleration functions are broadband, therefore most of structures with wide range of frequency of vibration could be affected by these acceleration functions. In a frequency range between 2.5 Hz and 10 Hz the average of PSD for ET acceleration functions and actual records are similar, therefore stochastic analysis of structures in those frequencies by ET acceleration functions and these ground motions may lead to comparatively similar results.

Stochastic analysis of structures with natural frequency less than 2.5 Hz and higher than 10 Hz by the ET method are more conservative as compared to real accelerograms; however they are similar for natural frequency between 2.5 Hz and 10 Hz. Such differences are to be expected because the design spectrum is not intended to match the response spectrum for any particular ground motion, but it is constructed to represent the average characteristics of a large set of ground motions with a margin of safety.

2.7 OTHER GROUND-MOTION PARAMETERS

A number of ground-motion parameters have been proposed to extract important information from each parameter.

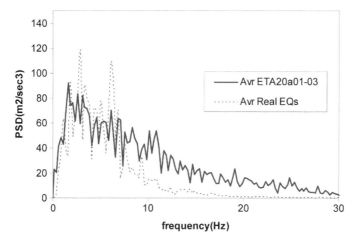

FIGURE 2.7 Comparison of Power Spectral Density for average of seven ground motions and average of ET acceleration functions at t = 10 sec.

2.7.1 ENERGY PARAMETERS

The energy spectrum may be used to provide additional important information about the damage potential of the earthquake ground motion related to these cumulative effects. Among energy parameters, input energy spectrum is directly related to ground motion as follows (Uang and Bertero 1988):

$$E_I = -\int m\ddot{v}_t \, dv_g \qquad (2.4)$$

Where $\ddot{v}_t$, is the absolute acceleration response of SDOF system, and v_g is ground displacement. In Figure 2.8(a) input energy for a system with $T = 0.7$ sec is depicted, and Figure 2.8(b) illustrates energy spectra for ET acceleration functions at $t = 10$ sec, and ground motions.

It is obvious that input energy for an ET acceleration function at $t = 10$ sec, does not conform to real accelerograms, especially at higher periods, and it is remarkably greater than that from ground motions. These observations led to the ideas for generating improved ETEFs in the next generations of these functions (Estekanchi et al. 2020)

Specific Energy Density (*SED*) is defined as:

$$SED = \int_0^{T_d} [\dot{v}_g(\tau)]^2 \, d\tau \qquad (2.5)$$

where $\dot{v}_g$, is ground motion velocity and T_d is duration of earthquake. *SED*s of records are illustrated in Figure 2.9.

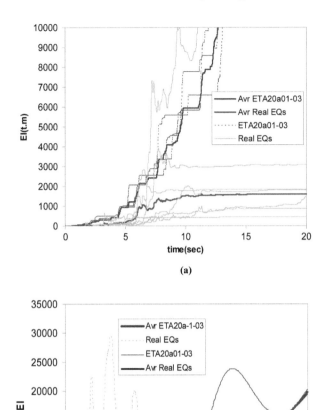

(a)

(b)

FIGURE 2.8   Comparison of input energy between ET acceleration functions and ground motions; (a) EI (Input Energy) at t = 0.7 sec, (b) energy spectra (t = 10 sec).

It is obvious that *SED* of ET acceleration functions is remarkably greater than average of earthquakes in such a way that *SED* of ET acceleration functions reach the average amount for selected ground motions at 2 seconds.

2.8   INTENSITY PARAMETERS

Arias Intensity (I_a) is closely related to root mean square of acceleration and is useful to characterize the frequency content and PSD of accelerograms.

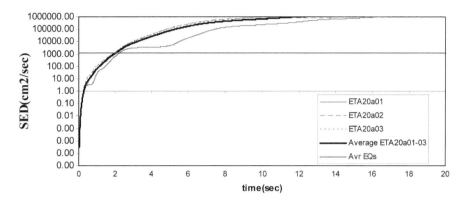

FIGURE 2.9 Comparison of SED between ET acceleration functions and ground motions.

$$I_a = \frac{\pi}{2g} \int_0^{td} [a_g(t)]^2 \, dt \tag{2.6}$$

The characteristic Intensity, I_c, is linearly related to an index of structural damage due to maximum deformations and absorbed hysteretic energy.

$$I_c = (a_{rms})^{3/2} \sqrt{T_d} \tag{2.7}$$

A95 is defined as that level of acceleration which contains up to 95 percent of the Arias Intensity (Sarma and Yang 1987).

Acceleration Spectrum Intensity (*ASI*) and Velocity Spectrum Intensity (*VSI*), are defined as

$$ASI = \int_{0.1}^{0.5} S_a \left(\xi = 0.05, T \right) dT \tag{2.8}$$

$$VSI = \int_{0.1}^{2.5} S_v \left(\xi = 0.05, T \right) dT \tag{2.9}$$

The above parameters are calculated for ETA20a acceleration functions and compared with the average of seven earthquakes. Results are depicted in Figure 2.10 through Figure 2.14.

As can be seen in Figure 2.10, Arias Intensity parameter for ETA20a acceleration functions matches the average of selected ground motions at about $t = 8$ sec, and increases with a hyperbolic trend with time.

As can be seen in Figure 2.11, A95 parameter is a nearly linear function of time and matches the average of selected earthquakes at about $t = 9$ sec. Also, from Figure 2.11, the parameter I_c of average of ground motions is equal to that of ET acceleration functions at $t = 6$ sec.

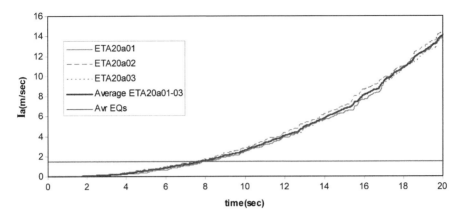

FIGURE 2.10   Ia for ETA20a acceleration functions and ground motions.

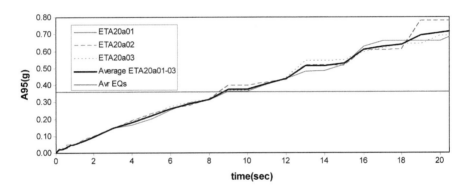

FIGURE 2.11   A95 for ET acceleration functions and ground motions.

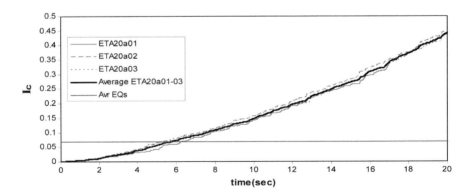

FIGURE 2.12   Ic for ET acceleration functions and ground motions.

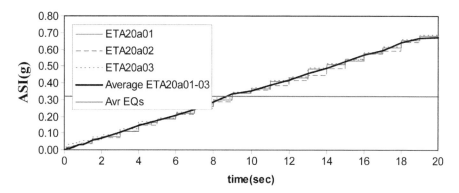

FIGURE 2.13   ASI for ET acceleration functions and ground motions.

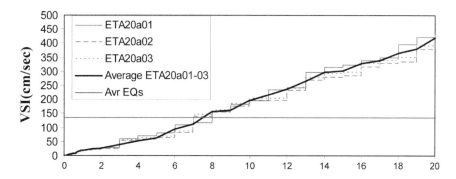

FIGURE 2.14   VSI for ET acceleration functions and ground motions.

Acceleration and velocity spectrum intensities are depicted and compared with ground motions in Figures 2.13 and 2.14.

It is evident from the figures that these two parameters linearly increase and match an average of ground motions at about $t = 8$ sec.

2.9   PERIOD PARAMETERS

The predominant period (T_p) is the period of vibration corresponding to the maximum value of the Fourier amplitude spectrum. It is seen from Figure 2.15, for ET acceleration functions this parameter varies between 0.2 second and 0.5 second, with the average of 0.35 second, which is equal to a predominant period of ground motions. This is due to same soil condition for earthquakes and ET acceleration functions.

Mean period (T_m) is defined as,

$$T_m = \frac{\sum C_i^2 / f_i}{\sum C_i^2} \qquad (2.10)$$

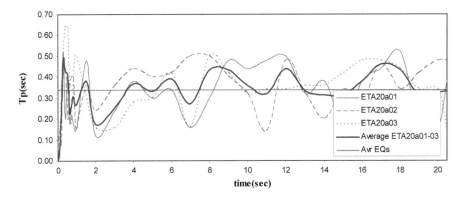

FIGURE 2.15   Tp of ET acceleration functions and ground motions.

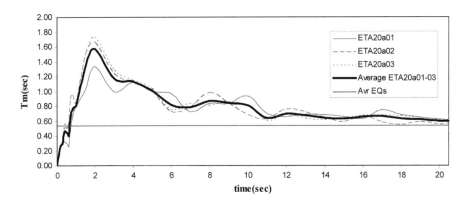

FIGURE 2.16   Tm of ET acceleration functions and ground motions.

where C_i is Fourier amplitude, and f_i represents the discrete Fourier transform frequencies between 0.25 Hz and 20 Hz. This is the best simplified frequency content characterization parameter.

Figure 2.16 shows that T_m, for ET acceleration functions are higher than ground motions.

2.10   PEAK VELOCITY ACCELERATION RATIO

This parameter reveals the dependence of magnitude and distance of earthquake from site. This ratio increases with increasing magnitude of earthquake and increasing source-to-site distance. This parameter is depicted in Figure 2.17.

It is evident that this value is very much higher for ETA20a acceleration functions than ground motions.

2.11   CUMULATIVE ABSOLUTE VELOCITY

The cumulative absolute velocity (CAV) is the area under absolute acceleration. This parameter correlates well with structural-damage potential (Kramer 1996).

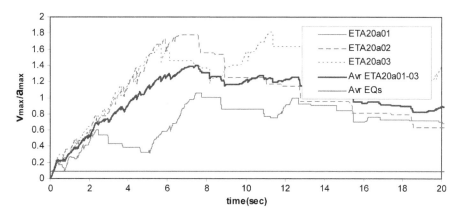

FIGURE 2.17   Vmax/amax for ET acceleration functions and ground motions.

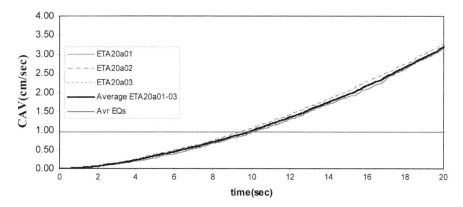

FIGURE 2.18   CAV for ET acceleration functions and ground motions.

From Figure 2.18, average of earthquakes and ET acceleration functions is the same at $t = 10$ sec. Therefore, some structural damages that are depended on CAV might be the same for selected earthquakes and ETA20a series of acceleration functions results at $t = 10$ sec.

2.12   SUSTAINED MAXIMUM ACCELERATION (SMA) AND VELOCITY (SMV)

SMA is defined as the third-highest absolute value of acceleration in the time history. These parameters are depicted in Figures 2.19 and 2.20.

It can be seen that these two parameters for ground motions are the same at $t = 6$ sec of ET acceleration functions.

2.13   EFFECTIVE DESIGN ACCELERATION (EDA)

EDA corresponds to the peak acceleration value found after low-pass filtering the input time history with a cut-off frequency of 9 Hz (Benjamin and Associates 1988).

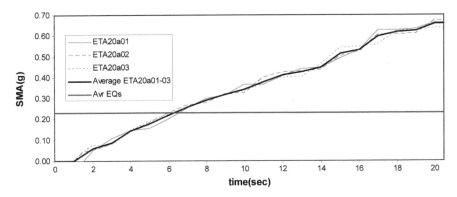

FIGURE 2.19   SMA for ET acceleration functions and ground motions.

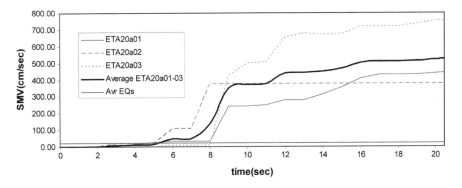

FIGURE 2.20   SMV for ET acceleration functions and ground motions.

Kennedy proposed that the EDA be 1.25 times of the third highest peak acceleration obtained from a filtered time history.

From Figure 2.21, EDA for the average of earthquakes is the same as ET acceleration functions at $t = 8.5$ sec.

2.14   SUMMARY AND CONCLUSIONS

In this chapter, characteristics of the first set of second-generation ET acceleration functions, that is, ETA20A01~3, as a set of synthesized intensifying accelerograms was investigated. These acceleration functions make use of a typical codified design spectrum as template spectrum. This study provides a template for extracting the characteristics of other ETEFs and comparing them to ground motion sets as well. The following conclusions can be drawn based on the results discussed in this research:

1. Most of the structurally significant parameters, except energy and amplitude parameters, which correspond to nonlinear behavior are nearly the same for

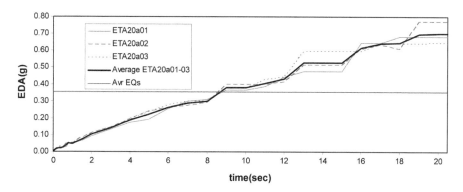

FIGURE 2.21 EDA for ET acceleration functions and ground motions.

ETA20a acceleration functions at the time of about $t = 10$ sec, (i.e., the target time) and average of ground motions at short periods.
2. At frequencies between 2.5 and 10 Hz, Fourier amplitudes of ETA20a are nearly the same as the average of selected ground motions. However, at high and very low frequencies, which are not covered in the ETA20a generation process (discussed in the next section), the differences are significant.
3. Intensity parameters for ETA20a acceleration functions around t = 8 sec, are comparable to average of selected ground motions.
4. At short periods, energy spectra of ETA20a acceleration functions at t = 10 sec, are similar to the average of selected earthquakes, however at middle and long periods they are not the same, and values from ETA20a acceleration functions are remarkably greater than ground motions.
5. ETA20a acceleration functions are in general not suitable for application in cases involving very short and very high effective periods of vibration. This includes highly nonlinear structures with a long period of vibration (above about 4 sec) on one side, and structures with a very short period of vibration (bellow about 0.1 sec) on the other side of the spectrum range. Newer and more advanced sets of ETEFs are available and should be used in practical applications.

NOTE

1 *Chapter Source*: Valamanesh, V., H.E. Estekanchi, and A. Vafai 2010. "Characteristics of Second Generation Endurance Time Accelerograms." *Scientia Iranica* 17, no. 1, pp. 53–61.

REFERENCES

Benjamin, J.R., and Associates. 1988. *A Criterion for Determining Exceedance of the Operating Basis Earthquake*. EPRI Report NP-5930. Palo Alto, CA: Electric Power Research Institute.
BHRC. 2005. *Iranian Code of Practice for Seismic Resistant Design of Buildings*, 3rd ed. Standard No. 2800-05. Tehran: Building and Housing Research Center.

Elenas, A. 2000 "Correlation Between Seismic Acceleration Parameters and Overall Structural Damage Indices of Buildings." *Soil Dynamics and Earthquake Engineering* 20, no. 1, pp. 93–100.

Estekanchi, H.E., K. Arjomandi, and A. Vafai. 2008. "Estimating Structural Damage of Steel Moment Frames by Endurance Time Method." *Journal of Constructional Steel Research* 64, no. 2, pp. 145–155.

Estekanchi, H.E., M. Mashayekhi, H. Vafai, G. Ahmadi, S.A. Mirfarhadi, and M. Harati. 2020, October. "A State-of-knowledge Review on the Endurance Time Method." *Structures* vol. 27, pp. 2288–2299.

Estekanchi, H., A. Vafai, and M. Sadeghazar. 2004. "Endurance Time Method for Seismic Analysis and Design of Structures." *Scientia Iranica* 11, no. 4, pp. 361–370.

Estekanchi, H.E., V. Valamanesh, and A. Vafai. 2007. "Application of Endurance Time Method in Linear Seismic Analysis." *Engineering Structures* 29, no. 10, pp. 2551–2562.

FEMA 356 2000 "NEHRP Guidelines for the Seismic Rehabilitation of Buildings." Federal Emergency Management Agency (FEMA): Washington, DC.

Kramer, S.L. 1996. *Geotechnical Earthquake Engineering*. Upper Saddle River, NJ: Prentice Hall.

Sarma, S.K., and K.S. Yang. 1987. "An Evaluation of Strong Motion Records and a New Parameter A95." *Earthquake Engineering and Structural Dynamics* 15, no. 1, pp. 119–132.

Uang, C.M., and V.V. Bertero. 1988. "Implications of Recorded Earthquake Ground Motions on Seismic Design of Building Structures." UCB/EERC-88/13. Berkeley: Earthquake Engineering Research Center, University of California.

3 Duration Properties of the Endurance Time Excitation Functions

3.1 INTRODUCTION

In Chapter 2, characteristics of a series of ETEFs were examined using some general parameters that are used to characterize the ground motions.[1] The earthquake motions should be characterized not only by using the parameters related primarily to amplitude of shaking, but also to the number of cycles and strong-motion duration which could play a role in the response they generate in some structures. There are many studies investigating the correlation between structural responses and parameter related to strong-motion duration. Their conclusions differ profoundly with respect to the influence of strong-motion duration on structural response. In their studies, Hancock and Bommer (2007) have concluded that duration is a secondary parameter and exploration for a direct correlation between duration and damage is not practical. In fact, strong-motion duration affects various types of damage indices in a different manner. Predominantly, damage indices related to cumulative energy and accumulated damage, such as absorbed hysteretic energy and fatigue, have a positive correlation with strong-motion duration; whereas damage indices related to the maximum response, such as the maximum interstory drift, do not have such a strong correlation. Moreover, Hancock and Bommer (2006) have asserted that this phenomenon is further dependent on the type of structural model that is used. Consequently, a structure with strength and stiffness degrading material is more sensitive to the number of cycles of motion and thus to strong-motion duration (Hancock and Bommer 2006). Despite the fact that more than forty definitions for strong-motion duration have been proposed by various researchers, there is no universally accepted approach to determine strong-motion duration.

In earthquake engineering, the dynamic response history analysis is recognized as a method that can incorporate almost all sorts of material and geometric complexities in a more realistic manner compared to the other methods used for structural analysis. As a result of these advantages, the tendency toward applying dynamic analysis has been increasing more rapidly in recent years. However, there are still a number of obstacles that prevent widespread use of this method. The Endurance Time (ET) method is a dynamic analysis procedure that uses intensifying accelerograms. In this method, the characteristics of the applied accelerograms constitute the predominant parameter that appreciably affects the reliability of the analysis results.

DOI: 10.1201/9781003217473-3

These accelerograms are basically generated so that they match to the target spectrum (such as code spectrum) at certain time, called target time, and remain proportional to it at all other times. In addition to amplitude parameters, strong ground-motion parameters should be considered in the generation process of these accelerograms as well. It is known that stronger earthquakes tend to have longer durations (Mashayekhi et al. 2020). While this observation tends to be consistent with the concept of the ET method, to keep the comparison as simple as possible only a single set of ground motions will be used in this chapter. The question that remains is which strong-motion definition is the most useful indicator of shaking characteristics of an earth-quake motion (Valamanesh and Estekanchi 2010). This chapter focuses on studying existing strong-motion duration definitions that have better correlation with the structural damage considering the ET analysis concepts. This information can be useful for proper application and also the generation of ET excitation functions.

3.2   GROUND MOTION SELECTION

Nonlinear dynamic analysis is a popular procedure for reliable seismic assessment of structural responses. In this procedure, the selection of an appropriate set of ground motions for dynamic loading is an important consideration because it can influence the response of structures. The typical design code recommendations for record selection are regarded as rather simplified considering the complicated procedures recommended for the selection process in the dynamic analysis. Most contemporary seismic codes, such as ASCE standards 7–16 (ASCE 2016), describe a relatively similar procedure for selection of seismic input motions to be used as dynamic loading in structures. Seismic motions can be either real or simulated records, while several important seismological parameters – such as magnitude, distance, and local site conditions – should reflect in a local seismic scenario (Katsanos, Sextos, and Manolis 2010). Whenever a set of accelerograms are selected on the basis of criteria such as Magnitude (M)-distance (R) pairs, a significant variability of the calculated response is observed. This issue is attributed to neglecting other significant parameters that should also be used to characterize the ground motions (Kappos and Kyriakakis 2000).

In this chapter, the record set proposed by the FEMA695 (Applied Technology Council 2009) for collapse assessment of building structure, is used. While the aforementioned record-selection procedures are mainly dependent on seismological conditions, the FEMA695's record set is selected so that they can be considered as applicable to more general site and source conditions.

3.3   ET EXCITATION FUNCTIONS

ET excitation functions (ETEFs) are intensifying accelerograms that make meaningful correspondence between the response of a structure at a particular time in ET analysis and the expected average of response to ground motions (Estekanchi et al. 2011, pp. 2535–46; Estekanchi et al. 2011, pp. 289–301; Estekanchi, Valamanesh, and Vafai 2007, pp. 2551–62). These ground motions may correspond to the seismicity of a particular site at certain hazard levels. Generally, the spectrum of the ET excitation functions can be at any time attributed to the spectrum associated with a particular hazard level. In order to simplify the interpretation, the spectrum of ET accelerogram can be considered only at target time to match a target spectrum. The target spectrum can be a code spectrum or an average spectrum of a set of ground motions. At other times, the produced spectrum by the ET excitation functions can be considered to vary linearly as:

$$S_{aC}(T,t) = \frac{t}{t_{target}} S_{aC}(T) \qquad (3.1)$$

where $S_{aC}(T)$ is the target spectrum, $S_{aC}(T,t)$ is the spectrum to be produced at time (t) by ET excitation functions, and t_{target} is the target time.

Displacement spectra are also a highly important consideration in characterizing a dynamic excitation. The target displacement spectrum can be defined as a function of the acceleration spectrum for linear analysis as (Estekanchi, Valamanesh, and Vafai 2007):

$$S_{uC}(T,t) = \frac{t}{t_{target}} S_{aC}(T) \times \frac{T^2}{4\pi^2} \qquad (3.2)$$

where $S_{uC}(T, t)$ is the target displacement spectrum to be induced at time t by the ET excitation functions.

In the second generation of ET excitation functions, the concepts of response spectrum and numerical optimization were used, and consistent results with conventional dynamic analysis were numerically achieved (Valamanesh and Estekanchi 2010). By extending the range of vibration period into long periods, the records in this generation also produced reasonable estimates in nonlinear range of behavior (Riahi and Estekanchi 2010). In the third generation, nonlinear response spectra were included in the optimization procedure (Nozari and Estekanchi 2011). In this chapter, four series of the ETEFs, which are presented in Table 3.1, are examined.

TABLE 3.1
Characteristics of Used ET Accelerograms

Series	Target spectrum	Long periods covered	Nonlinear optimization
ETA20a	Code spectrum (standard 2800)*	N	N
ETA20e	Average of several recorded motion on stiff soil	Y	N
ETA40g	Code spectrum (ASCE standard)	Y	N
ETA20en	Average of several recorded motion on stiff soil	-	Y

Note: * Iranian national code.

3.4   A REVIEW OF DEFINITIONS OF STRONG-MOTION DURATION

Strong-motion duration definitions can be classified into three generic groups, Bracketed Duration, Uniform Duration, and Significant Duration. The Bracketed Duration, D_b, is defined as the total time of a motion that elapsed between the first and last extrusion of a specified level of acceleration, a_0 (Bommer and Martinez-Pereira 1999), as schematically depicted in Figure 3.1 for an accelerogram using threshold of 0.05g.

The second group is Uniform Durations, D_U, defined as the sum of time intervals during which the acceleration is greater than the specified threshold (Bommer and Martinez-Pereira 1999), as illustrated in Figure 3.2.

The third group is called Significant Duration, D_s, and these are based on accumulation of energy in an accelerogram represented by the integral of the square-of-the-ground acceleration, velocity, and displacement. If the integral of the square-of-the-ground acceleration is employed, the quantity is related to the Arias intensity, AI (Arias 1970).

$$AI = \frac{\pi}{2g} \int_0^{t_r} a^2(t) dt \tag{3.3}$$

Here, t_r is defined as total duration of the accelerogram, $a(t)$ is the acceleration time history, and g is the acceleration due to the gravity. Significant Duration is defined as the time interval over which some specified proportion of the total energy is accumulated (Bommer and Martinez-Pereira 1999). This approach for limits of 10 to 90 percent of the total energy for an accelerogram is illustrated on a plot of the build-up of AI, in Figure 3.3.

The root-mean-square of an accelerogram is defined as (Bommer and Martinez-Pereira 1999):

$$a_{rms} = \frac{1}{t_2 - t_1} \int_{t_1}^{t_2} a^2(t) dt \tag{3.4}$$

where t_1 and t_2 are beginning and end of the time interval under consideration, respectively. Any definition based on root-mean-square of an accelerogram is categorized into Significant Duration (Bommer and Martinez-Pereira 1999).

McCann and Shah (1979) have defined strong-motion duration by plotting the cumulative a_{rms} of the accelerogram, noting that beyond a certain point, it begins to decay. The end of the strong-motion phase is determined by plotting the derivate of cumulative a_{rms} function against time and getting the time beyond which a_{rms} remains negative (Ibarra, Medina, and Krawinkler 2005). The start of the strong-motion phase is determined in exactly the same approach by using the reverse acceleration time-history (Bommer and Martinez-Pereira 1999). This procedure is depicted in Figure 3.4.

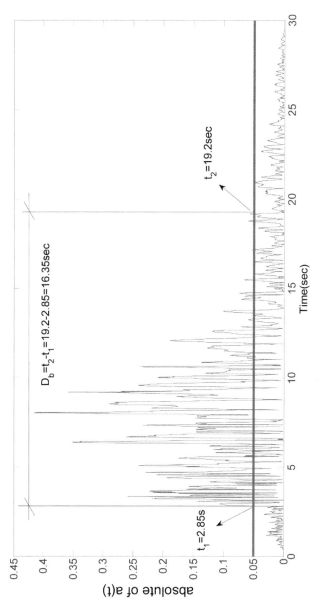

FIGURE 3.1 "Bracketed duration" of an accelerogram.

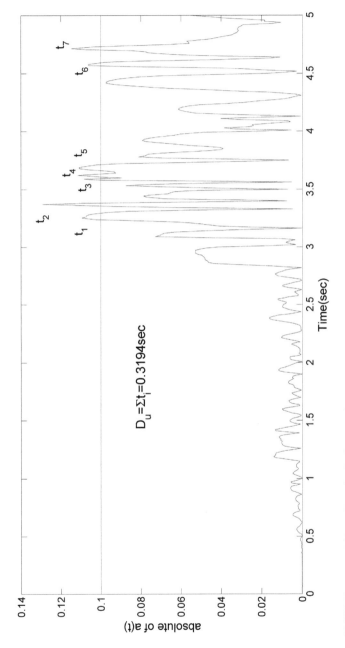

FIGURE 3.2 "Uniform Duration" of an accelerogram.

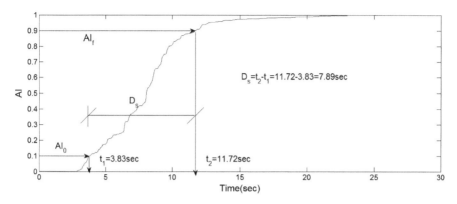

FIGURE 3.3 "Significant Duration" of an accelerogram.

3.5 COMPARISON BETWEEN ET ACCELEROGRAMS AND REAL GROUND MOTIONS

While ETEFs are theoretically an infinitely increasing excitation, duration definitions can be applied to them if a certain window of them, from $t = 0$ to $t = t_I$, is considered. The strong-motion duration of both the ET records and real-ground motions using different duration definitions are calculated in this chapter. The ET records at each time window can be considered as a single motion; for instance, 10-second windows or 20-second windows of an ET record are two separate motions. Therefore, an ET record is not inherently a single motion and, hence, its strong-motion duration is varied against time. It is noteworthy that, unlike Significant Duration, Bracketed Duration and Uniform Duration depend on how the record is scaled. In this study, the ET records and the real-ground motions are scaled to EPA (Effective Peak Acceleration) of 0.35g. For the ET records, a scaling process is performed for each window. Afterwards, the target time is identified as the time in which motion duration of the ET records will be equal to the ones associated with the real-ground motions. The target time is determined schematically and presented in Figure 3.5.

A similar procedure is performed for other series of the ET records and different strong-motion duration definitions. The target times associated with different series of the ET accelerograms and different strong-motion definitions are presented in Table 3.2. Dispersion of the results is evident.

3.6 EVALUATION OF PROPOSED TARGET TIME

In order to evaluate the effectiveness of each target time presented in Table 3.2, several degrading models with different periods and different ductility ratio are constructed and then subjected to both the ET records and the real-ground motions. It should be noted that the ET records are scaled to the proposed target times which are computed using different strong-motion definitions. In this study, the peak-oriented model is employed to characterize the hysteretic behavior of materials. This model keeps the basic hysteretic rules proposed by Clough and Johnston (1966) and later modified by

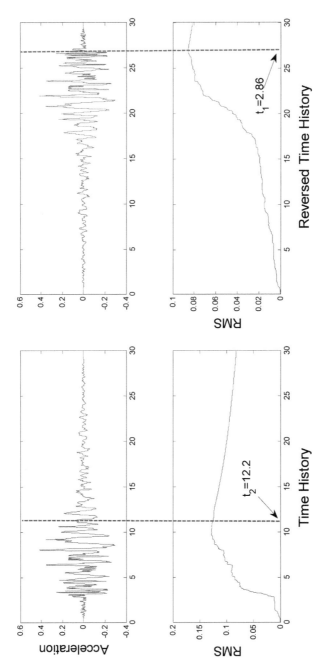

FIGURE 3.4   Definition of strong-motion duration by McCann and Shah (1979).

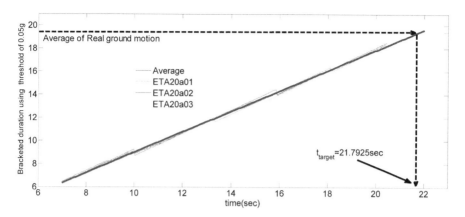

FIGURE 3.5 Procedure to determine the target time.

TABLE 3.2
Target Time for Different Series of ET Accelerogram

	Target time						
	D_b (0.05g)*	D_b (0.1g)	D_u (0.05g)	D_u (0.01g)	RMS	D_s (5.75%)	D_s (5.95%)
ETA20a	25.29	14.64	11.12	3.84	22.49	9.63	19.11
ETA20e	27.09	15.65	12.84	4.52	23.41	9.81	20.56
ETA40g	26.61	15.29	13.03	4.00	22.59	9.53	20.54
ETA20en	28.08	15.35	13.17	4.13	22.69	9.68	20.51

Note: * Quantity in the parenthesis indicates the thresholds used in the duration definitions.

Mahin and Bertero (1976), but the backbone curve is modified to include strength capping and residual strength (Ibarra, Medina, and Krawinkler 2005). A basic rule for a peak-oriented hysteretic model is illustrated in Figure 3.6.

Moreover, a Damage Index proposed by Kunnath and Jenne (1994) is considered as a damage indicator in this study. Kunnath, Reinhorn, and Lobo (1992) modified the Park-Ang damage index as:

$$\bar{D}_{PK}{}^K = \frac{\mu_{\theta m} - \mu_{\theta y}}{\mu_{\theta u} - \mu_{\theta y}} + \bar{\beta}_K \frac{\overline{E_{hm}}}{M_y \mu_{\theta u} \theta_y} \tag{3.5}$$

where $\mu_{\theta u}$ is the ultimate rotation ductility under a monotonic static load. Parameters used for characterizing the hysteretic behavior of material are: $\alpha_1 = 0.1$, $\alpha_2 = -0.03$, $\frac{\theta_c}{\theta_y} = 11$, $\mu_{\theta u} = 8$, $\bar{\beta}_K = 0.15$ where α_1, α_2 are the post yielding stiffness ratio and

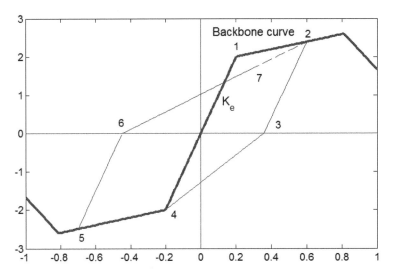

FIGURE 3.6   Basic rules for peak-oriented hysteretic model.

the post capping stiffness ratio, respectively. A high value of $\overline{\beta_K}$ used in this study implies the higher contribution of hysteretic energy dissipation to damage. The latter statement guarantees that the used damage index accounts for the duration of strong-motion and cumulative-inelastic action. The assumed range of periods of interest is 0.2 to 3 seconds for all motions, and the range of ductility is 2 to 6.

Afterwards, the designed structures are subjected to both the ET records and real-ground motions. Both the ET records and real-ground motions are scaled. The scaling process of real-ground motion records only considers the spectral value at the natural vibration period of the structures. On the other hand, the scaling process of ET records further considers the time at which the ET records reach the target spectrum, target time. In this study, the target time of ET records considering different strong-motion definitions is determined so that the ET records produce consistent duration compared to real-ground motions.

For comparison, the damage indices of structures when subjected to the ET records are plotted versus the ones when subjected to ground motions. Also, for the quantitative comparison of each proposed target time, δ parameter is defined as:

$$\delta = \sqrt{\frac{1}{N}\sum_{i=1}^{N}\left(\frac{DI_{RG} - DI_{ET}}{DI_{RG}}\right)^2} \qquad (3.6)$$

where DI_{RG} and DI_{ET} are damage indices of structure when they are subjected to real ground motions and the ET records, respectively. The number N is the number of structures that are considered in this study. Table 3.3 represents this value for

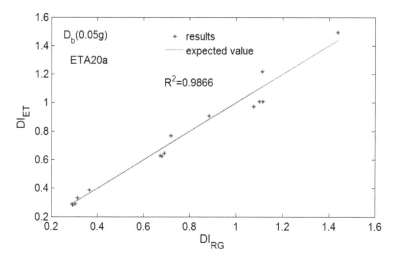

FIGURE 3.7   Revelation of correlation between the results of ET records and real ground motions.

TABLE 3.3
δ Value for Different Series of ET Accelerogram and Different Strong-Motion Definitions

	δ value						
Series	**D_b (0.05g)**	**D_b (0.1g)**	**D_u (0.05g)**	**D_u (0.1g)**	**RMS**	**D_s (5.75%)**	**D_s (5–95%)**
ETA20a	-	0.017	0.030	0.059	-	0.030	0.019
ETA20e	-	0.035	0.229	0.300	-	0.011	0.957
ETA40g	0.049	0.027	0.019	0.034	0.038	0.022	0.048
ETA20en	-	0.030	0.275	0.507	-	0.026	0.906

Note: (-) line indicates that the mentioned value cannot be calculated because there is no record with that duration.

different motion duration definitions and different series of ET accelerograms. In addition, linear correlation factor for data is calculated. Table 3.4 represents this value for different series of ET accelerograms and different motion duration definitions. Figure 3.7 shows a high correlation between results of ET accelerograms and real-ground motions.

Table 3.3 displays a more effective scaling process for each series of ET accelerograms; for instance, scaling based on Bracketed Duration with threshold of 0.1g will be more rewarding for ETA20a. Table 3.4 shows the high correlation between results of ET accelerograms and real ground motions. Table 3.5, represents the comprehensive conclusion of above tables.

TABLE 3.4
Linear Correlation Factor for Different Series of ET Accelerograms and Different Strong-Motion Definitions

	R^2						
Series	D_b (0.05g)	D_b (0.1g)	D_u (0.05g)	D_u (0.1g)	RSM	D_s (5–75%)	D_s (5–95%)
ETA20a	-	0.987	0.998	0.976	-	0.993	0.990
ETA20e	-	0.991	0.992	0.995	-	0.996	0.943
ETA40g	0.981	0.975	0.990	0.995	0.978	0.997	0.962
ETA20en	-	0.992	0.991	0.992	-	0.975	0.962

Note: (-) line indicates that the mentioned value cannot be calculated because there is no record with that duration.

TABLE 3.5
Rankings of Different Series of ET Accelerograms Considering Different Criteria

Series	Best indicator	Target time	Lowest delta	Rank of lowest delta	Rank of highest R2
ETA20a	D_b (0.1g)	14.64	0.017	2	1
ETA20e	D_s (5–75%)	9.81	0.011	1	3
ETA40g	D_u (0.05g)	13.03	0.019	3	2
ETA20en	D_s (5–75%)	9.68	0.026	4	4

3.7   SUMMARY AND CONCLUSIONS

In this chapter, the concept and implementation of strong-motion duration in the ET analysis has been presented. Theoretically, ETEFs are infinitely intensifying functions. However, application of ground-motion duration to the particular windows of these functions can provide some insight into the characteristics of these functions as compared to ground motions. It should be noted that the duration of ground motions tends to increase with earthquake magnitude. However, in this chapter only a single set of ground motions at an effectively constant average duration has been considered. As a result of these studies, the following can be concluded:

1. Duration of motion does not differ widely for different series of ETEfs. It means that duration of the current ETEFs is not very sensitive to the target spectrum used in generating them.
2. The scaling process, which is more effective in making the best level of compatibility between the ET records and real ground motion, is not the same for different series of ET accelerograms; therefore, a single procedure for scaling ET accelerograms that is consistent with real motions cannot be prescribed.

3. The optimum target time that makes the best consistency between ET accelerograms and real ground motions differs widely for different series of ET accelerograms. This reveals that a single target time cannot provide the best level of possible consistency.

4. The high correlation between the results of ET accelerograms and real-ground motions reveals the acceptable performance of the ET records in the non-linear region. Contrary to what it was expected, the ETA20e series of ET accelerograms that are modified in the nonlinear region has relatively the weakest performance among the different series of ET accelerograms.

5. In the nonlinear region, the response of the SDOF structure not only depends on its natural period, but higher periods also can affect the response. This can be attributed to the elongation of the effective period of vibration as the nonlinear behavior of the structure becomes more dominant. As expected, the response of a structure subjected to ETA20a series, ETA20e series, and ETA40g series have similar generation approaches, but the target spectrum will be different due to the contribution of other periods. The ETA20e series of ET accelerograms are matched to the average spectrum of recorded motions on stiff soil. This similarity between spectral acceleration of the ET records and real-ground motions makes ETA20e series more duration-consistent compared to others in this regard.

NOTE

1 *Chapter Source*: Mashayekhi, M., and H.E. Estekanchi. 2013 "Investigation of Strong-Motion Duration Consistency in Endurance Time Excitation Functions." *Scientia Iranica* 20, no. 4, pp. 1085–1093.

REFERENCES

Applied Technology Council, 2009. Quantification of Building Seismic Performance Factors. FEMA, p. 695, US Department of Homeland Security.

Arias, A. 1970. "Measure of Earthquake Intensity." In *Seismic Design for Nuclear Power Plants*, ed. Robert J. Hansen. Cambridge MA: MIT Press, 439–483.

ASCE 2016. *Minimum Design Load for Building and Other Structures*. ASCE Standard No. 07-16, American Society of Civil Engineers.

Bommer, J.J., and A. Martinez-Pereira. 1999. "The Effective Duration of Earthquake Strong Motion." *Journal of Earthquake Engineering* 3, no. 2, pp. 137–172.

Clough, R.W., and S.B. Johnston. 1966. "Effect of Stiffness Degradation on Earthquake Ductility Requirements." *Proceeding of Japan Earthquake Engineering Symposium*. Tokyo.

Estekanchi, H.E., H.T. Riahi, and A. Vafai. 2011. "Application of Endurance Time Method in Seismic Assessment of Steel Frames." *Engineering Structures* 33, no. 9, pp. 2535–2546.

Estekanchi, H.E., A. Vafai, and V. Valamanesh. 2011. *Recent Advances in Seismic Assessment of Structures by Endurance Time Method*. Proceedings of a U.S.-Iran-Turkey Seismic Workshop – Seismic Risk Management in Urban Areas. PEER report 2011/07, pp. 289–301. December 14–16, 2010, Istanbul.

Estekanchi, H.E., V. Valamanesh, and A. Vafai. 2007. "Application of Endurance Time Method in Linear Seismic Analysis." *Engineering Structures* 29, no. 10, pp. 2551–2562.

Hancock, J., and J.J. Bommer. 2006. "A State of Knowledge Review of the Influence of Strong-Motion Duration on Structural Damage." *Earthquake Spectra* 22, no. 3, pp. 827–845.

Hancock, J., and J.J. Bommer. 2007. "Using Spectral Matched Records to Explore the Influence of Strong-Motion Duration on Inelastic Structural Response." *Soil Dynamics and Earthquake Engineering* 27, no. 4, pp. 291–299.

Ibarra, L.F., R.A. Medina, and H. Krawinkler. 2005. "Hysteretic Models that Incorporate Strength and Stiffness Deterioration." *Earthquake Engineering and Structural Dynamics* 34, no. 12, pp. 1489–1511.

Kappos, A.J., and P. Kyriakakis. 2000. "A Re-evaluation of Scaling Techniques for Natural Records." *Soil Dynamics and Earthquake Engineering* 20, no. 1, pp. 111–123.

Katsanos, E.I., A.G. Sextos, and G.D. Manolis. 2010. "Selection of Earthquake Ground Motion Records: A State-of-the-Art Review from a Structural Engineering Perspective." *Soil Dynamic and Earthquake Engineering* 30, no. 4, pp. 157–169.

Kunnath, S.K., and C. Jenne. 1994. "Seismic Damage Assessment of Inelastic RC Structures." *Proceedings of the 5th U.S. National Conference on Earthquake Engineering* 1, pp. 55–64. EERI, Oakland, CA.

Kunnath, S.K., A.M. Reinhorn, and R.F. Lobo. 1992. "IDARC Version 3.0: A Program for the Inelastic Damage Analysis of Reinforced Concrete Structures." Buffalo, NY: National Center for Earthquake Engineering Research.

Mahin, S.A., and V.V. Bertero. 1976. "Nonlinear Seismic Response of Coupled Wall System." *ASCE Journal of the Structural Division* 102, pp. 1759–1980.

Mashayekhi, M., M. Harati, A. Darzi, and H.E. Estekanchi. 2020. "Incorporation of Strong Motion Duration in Incremental-based Seismic Assessments." *Engineering Structures* 223, p. 111144.

McCann, M.W., and H.C. Shah. 1979. "Determining Strong-Motion Duration of Earthquakes." *Bulletin of the Seismological Society of America* 69, no. 4, pp. 1253–1256.

Nozari, A., and H.E. Estekanchi. 2011. "Optimization of Endurance Time Acceleration Functions for Seismic Assessment of Structures." *International Journal of Optimization in Civil Engineering* 1, no. 2, pp. 257–277.

Riahi, H.T., and H.E. Estekanchi. 2010. "Seismic Assessment of Steel Frames with Endurance Time Method." *Journal of Construction Steel Research* 66, no. 6, pp. 780–792.

Valamanesh, V., and H.E. Estekanchi. 2010. "Characteristics of Second Generation Endurance Time Method Accelerograms." *Scientia Iranica* 17, no. 1, pp. 53–61.

4 Generating ET Excitation Functions by Numerical Optimization

4.1 INTRODUCTION

In the ET method, the computational demand associated with response-history analysis is considerably reduced by subjecting the structure to an intensifying excitation function (ETEF) and monitoring the objective performance indexes through time.[1] Afterwards, structural performance can be evaluated based on the response of the system at each excitation level (Estekanchi, Valamanesh, and Vafai 2007; Estekanchi and Basim 2011). Generating appropriate dynamic inputs is essential for the ET method's success. With respect to this issue, an input function can be considered as appropriate if the analysis results estimated in the ET analysis are consistent with those under real earthquakes. The ETEFs currently applied in the ET method have two basic properties: (1) these functions are intensifying as their amplitude increases with time; (2) these functions are optimized such that the response spectrum of any window from $t = 0$ to $t = t1$ is proportional to a template response spectrum with a scale factor that increases with time (Estekanchi, Valamanesh, and Vafai 2007). As will be explained, generating the ETEFs with these properties is a formidably complicated problem from an analytical viewpoint; consequently, numerical optimization turns out to be a very viable approach in order to tackle this issue for now.

A general procedure for generating the ET acceleration functions (ETEFs) is illustrated in Figure 4.1. To generate these functions, a template response spectrum matching either a required design spectrum or a spectrum from a set of ground motions should be considered. Currently, different template spectrums have been considered. These include the design spectrum of the Iranian National Building Code (INBC), American Society of Civil Engineers (ASCE) design spectrum and average of response spectra from sets of ground motions (BHRC 2005; ASCE7-05 2005). Several sets of the ETEFs have been produced for each target response. A set of the ETEFs (ETA20d01-3) was optimized utilizing the INBC design spectrum as the target response spectrum. In this study, the optimization process of this set of the ETEFs is explained, and different approaches are proposed to improve these functions and to achieve more consistent results with the template design spectrum.

DOI: 10.1201/9781003217473-4

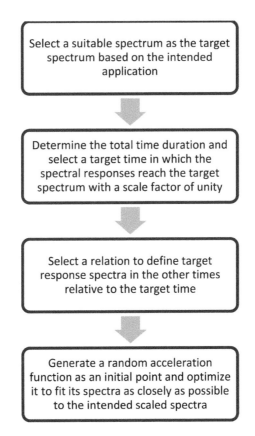

FIGURE 4.1   General procedure of generating ET acceleration functions.

4.2   GENERATING ET ACCELERATION FUNCTIONS

The ETA20d series of the ETEFs was generated utilizing design spectrum of standard No. 2800 of INBC for soil type (II) as the target response (Estekanchi, Valamanesh, and Vafai 2007; BHRC 2005). Duration of these ETEFs is 20.47 seconds, which consists of 2,048 acceleration points in 0.01-second time steps. The target time of the functions is the 10th second when the response of a Single Degree of Freedom (SDOF) system with a damping ratio of 5 percent equals the codified template design spectrum with a scale factor of unity. Objective response in all other times is defined by a linear function of time based on the target response as follows:

$$S_{aT}\left(T_i, t_j\right) = \frac{t_j}{t_{Target}} S_{aC}\left(T_i\right)$$

$$S_{i,j}^{aT} = S_{aT}\left(T_i, t_j\right)$$

(4.1)

where, T is the fundamental period of structure, t_{Target} is the target time, S_{aT} is the target acceleration response of structure, and S_{aC} is the codified spectral acceleration, which can be obtained from Equation 4.2 below:

$$
\begin{cases}
B = 1 + 1.5\left(\dfrac{T}{0.1}\right) & T < 0.1s \\[2mm]
B = 2.5 & 0.1s \leq T < 0.5s \\[2mm]
B = 2.5(\dfrac{0.5}{T})^{\frac{2}{3}} & 0.5s \leq T
\end{cases}
\tag{4.2}
$$

$$
S_{aC} = \frac{0.35BI}{R}
$$

Here I is the importance factor of under design building considered to be 1.0, and R is the response reduction factor that has not been applied (i.e., assumed equal to 1.0) (BHRC 2005).

Similarly, the displacement target response can be obtained from the codified spectral acceleration using basic dynamic properties as follows:

$$
S_{uT}\left(T_i, t_j\right) = \frac{t_j}{t_{Target}} S_{aC}\left(T_i\right) \times \frac{T_i^2}{4\pi^2}
$$
$$
S_{i,j}^{uT} = S_{uT}\left(T_i, t_j\right)
\tag{4.3}
$$

Since the 10th second is selected as the target time, it is obvious that the target response, for example, at the 5th second, is half of the codified value, and at the 20th second is twice the codified value. The objective response will be in m*n matrix form, where the number of rows (m) is equal to the number of period points and the number of columns (n) is equal to the number of time steps. Thus, t_j can be formulated as follows:

$$
t_j = j \times dt, \qquad j = 1, 2, \ldots, n
\tag{4.4}
$$

For calculation of time-history responses due to a dynamic input, one can consider the differential equation of motion for an SDOF system under an earthquake excitation:

$$
\ddot{u}(t) + 2\xi\omega_n \dot{u}(t) + \omega_n^2 u(t) = -\ddot{u}_g(t)
\tag{4.5}
$$

Here, ξ is the damping ratio, ω_n is the natural circular frequency which corresponds to the natural period of vibration with $2\pi / T$, and $\ddot{u}_g(t)$ is the ground excitation time history. The acceleration response function can be calculated from absolute acceleration responses as follows:

$$S_a\left(T_i,t_j\right) = \max\left(\left\|\ddot{u}(\tau)+\ddot{u}_g(\tau)\right\|\right) \qquad 0 \le \tau \le t_j$$
$$S_{i,j}^a = S_a\left(T_i,t_j\right) \tag{4.6}$$

Further the displacement response function can be obtained from relative displacement responses as:

$$S_u\left(T_i,t_j\right) = \max\left(\left\|u(\tau)\right\|\right) \qquad 0 \le \tau \le t_j$$
$$S_{i,j}^u = S_u\left(T_i,t_j\right) \tag{4.7}$$

Now, the problem is approached by formulating it as an unconstrained optimization problem in the time domain with the following objective function, which can be minimized using numerical methods:

$$F\left(a_g\right) = \sqrt{\frac{\sum\limits_{i=1}^{n}\sum\limits_{j=1}^{m}\left\{\left[S_{i,j}^a - S_{i,j}^{aT}\right]^2 + \alpha\left[S_{i,j}^u - S_{i,j}^{uT}\right]^2\right\}}{n \times m}} \tag{4.8}$$

Where $a_g(t)$ consists of the ETEF acceleration points as the optimization variables.

It should be noted that either acceleration or displacement response or a combination of them (or other criteria) can be utilized as the target response (Mashayekhi, Estekanchi, and Vafai 2020). Since the acceleration and displacement responses are closely correlated for an SDOF system, only one of them may also be considered as the target response (Estekanchi, Valamanesh, and Vafai 2007; Riahi and Estekanchi 2010). For simplicity of explanation, acceleration response alone is selected as the objective function of optimization here (i.e., the weight parameter α assumed equal to 0).

Different optimization approaches can be adopted for solving the problem (Mashayekhiet et al. 2019). As a simple setup for the optimization process, an unconstrained optimization procedure that applies a quasi-Newton algorithm is used here (Nozari and Estekanchi 2011). Two hundred period points are distributed logarithmically in the range of 0 and 5 seconds, and 20 long-period points are used to control displacements. In addition, the damping ratio is assumed to be 5 percent for all of the SDOF systems. Since the structural responses are calculated in all the time steps, the response function is produced in a 220*2048 matrix, which needs 220 time-history analyses for its calculation in each cycle.

A typical ET acceleration function generated utilizing this approach is shown in Figure 4.2. The acceleration and displacement response spectra of these functions at the 5th, 10th, 15th, and 20th seconds are illustrated in Figure 4.3. In Figure 4.4, the acceleration and displacement response spectra and average response spectra of the three ETEFs can be seen at the target time matching the template spectrum with a

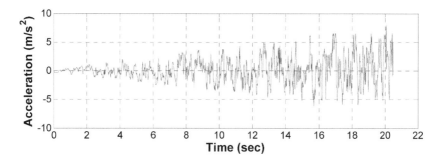

FIGURE 4.2 ETA20d01 acceleration function.

scale factor of unity. As can be seen, the optimization procedure has been successful in producing the ETEFs that are matched with the specified target with reasonable accuracy. These ETEFs are available online (Estekanchi 2020) as well. Dynamic properties of the ETEFs produced employing this procedure are investigated by Valamanesh and colleagues (2010).

The calculated errors for each ETEF are given in Table 4.1. Two approaches can be applied for the error calculation: the first approach is the same as the objective function for the optimization process, which is characterized by Relation 8, and the second one, the so-called base error, is similar to the first approach; however, by definition, its purpose is to negate the effect of period points distribution and the optimization time steps in calculation of errors. To calculate the base error, period points between 0 to 5 seconds with uniform distribution, with a step size of 0.005 seconds, and all of the time steps (2,048) are considered. Therefore, the function for calculating the base error will be a 1001×2048 matrix. The base error is applied in order to compare the convergence of the different ETEFs to the target of a perfect match with the optimization objective.

As can be seen, the average of base errors for the three ETEFs of the ETA20d series is 0.6284 m/s^2, and the error of average response of these ETEFs is 0.4564 m/s^2. Thus, by averaging the results from the three records, the amount of deviation is reduced about 27 percent. Preliminary studies have shown that applying a set of three ETEFs is effective in decreasing the average response error, and the advantage of accuracy diminishes because of the required additional computations with increasing the number of ETEFs to more than three (Estekanchi, Valamanesh, and Vafai 2007; Nozari and Estekanchi 2011). Thus, averaging the results of the three ET analyses is recommended in ordinary ET analyses as a balanced solution to minimize calculations while reducing in the results the effect of random scattering and obtaining a sense of the expected level of dispersion in the estimates.

The optimization of ETA20d series of the ETEFs using the unconstrained optimization procedure, considering the high volume of computations, is a time-consuming process – for example, to produce each ETEF, more than 120 hours was required by utilizing Pentium IV CPU with a frequency of 2,800 GHz. Therefore, improving the optimization procedure is essential to make the process practically more appealing.

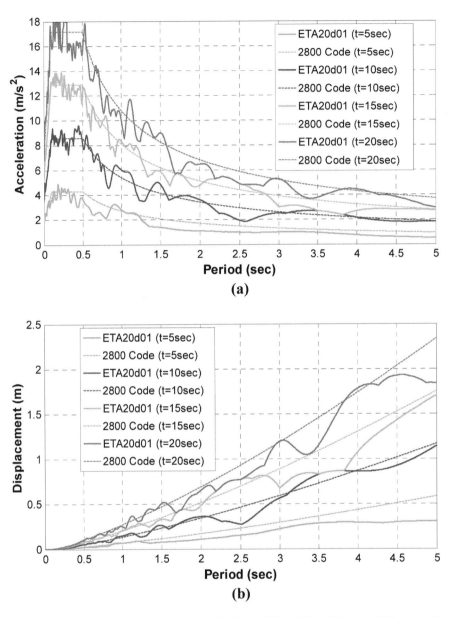

FIGURE 4.3 Response spectra of ETA20d01 at 5th, 10th, 15th, and 20th seconds, (a) acceleration, (b) displacement.

4.3   NONLINEAR LEAST SQUARES FORMULATION

As was shown, to produce the ETA20d series of the ETEFs, the objective function of optimization is formulated as the square root of the sum of squares. However, the objective function of optimization can further be defined in the form of the

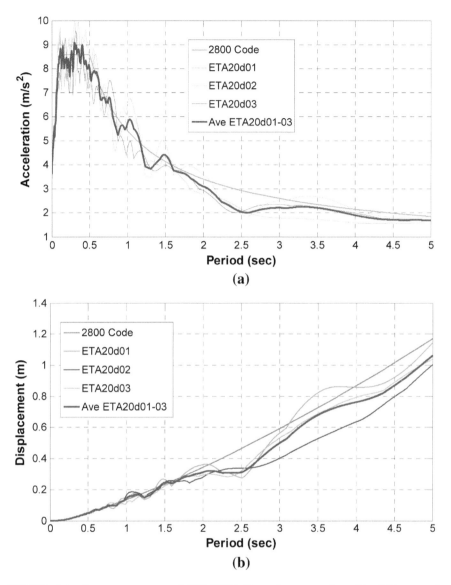

FIGURE 4.4 Response spectra of ETA20d01-03 at the target time (the 10th second), (a) acceleration, (b) displacement.

least squares, and special algorithms for the optimization of nonlinear least squares problems can be applied. By employing this method, a computer code was developed that takes the objective function in a matrix form and proceeds to minimize every element in the matrix, utilizing the two different algorithms for two functional states:

1. If the number of elements in the objective matrix of optimization is fewer than the number of optimization variables, the program will use a quasi-Newton

TABLE 4.1

Errors of Acceleration Responses of the First Series ETEFs

	Absolute error (m/s²)	
Acceleration function	Optimized points	Base points
ETA20d01	0.5094	0.5378
ETA20d02	0.7126	0.7929
ETA20d03	0.5321	0.5545
Average	0.5847	0.6284
Ave ETA20d01-03	0.4141	0.4564

algorithm similar to the procedure used in previous unconstrained optimization (Nozari and Estekanchi 2011).

2. If the number of elements in the objective function is equal to or higher than the number of optimization variables, the program will utilize the Trust Region algorithm based on the Interior Reflective Newton method, which is optimized for nonlinear least squares problems (Coleman and Li 1996; Coleman and Li 1994).

In fact, it is important to consider the higher power of the Trust Region algorithm in the second state in order to optimize the least squares problems, and what is important is the transformation of the objective function into an appropriate form for this function. Hence, the objective function of optimization is expressed as follows:

$$M_{i,j}\left(a_g\right) = \left\{\left[S_{i,j}^a - S_{i,j}^{aT}\right] + \alpha\left[S_{i,j}^u - S_{i,j}^{uT}\right]\right\} \quad (4.9)$$

In some cases, computer or optimization software memory may not be capable of handling too-large matrices in optimization. For example, in the current case, after performing a number of preliminary trials and errors, it was concluded that the available memory of the computer was not sufficient to define the complete ET objective function. For a problem with 2,048 variables, an objective function in a matrix form with at most 9,000 elements could be defined (Nozari and Estekanchi 2011). In these cases, in order to utilize the program the objective function needs to be compressed in such a way that it conforms to the limitations of the usable memory. The idea is to apply a discrete time definition for the objective function, such that a limited number of identical times are chosen for the optimization:

$$t_k = p \times k \times dt, \quad k = 1, 2, ..., l, \quad l = \left[\frac{n}{p}\right] \quad (4.10)$$

where, t_k is the discrete times included in the objective function, p is an integer parameter to determine the discrete time intervals, and dt is the time-history analysis time step (assumed equal to 0.01 seconds).

As a result, a smaller-sized matrix can be produced from the initial objective matrix. It should be mentioned that, considering the concept of the ETEFs and the definition of spectral responses, responses at the times between two consecutive discrete times are restricted within the responses at those discrete times:

$$t_k \leq t_j \leq t_{k+1} \Rightarrow \begin{cases} S^u_{i,k} \leq S^u_{i,j} \leq S^u_{i,k+1} \\ S^a_{i,k} \leq S^a_{i,j} \leq S^a_{i,k+1} \end{cases}$$

(4.11)

In fact, by utilizing this procedure, we can avoid changing the time step of time-history analysis and instead retain the time step of 0.01 seconds, and the results accuracy is not affected, while the size of the objective function is reduced. Therefore, an objective function similar to that used in the production of ETA20d series is formed. While it retains its matrix form, only specific time steps are chosen and all considered period points (all the columns of the objective matrix) are retained, resulting in an objective function with smaller size.

4.4 IMPROVED ET ACCELERATION FUNCTIONS

As a result of utilizing the new formulation of discrete ET objective function, a new series of the ETEFs is produced. For the optimization of these ETEFs, the dynamic properties of former functions, presented in Chapter 2, are utilized. To consider the computational limitations explained above, after performing the primary experiments, selective times are considered to begin at 0.5 seconds with equal intervals of 0.5 seconds, ending at the 20.00 seconds duration (i.e., p assumed equal to 50 so l equals to 40). Therefore, taking the 220 points of the period into account, the initial 220×2048 objective matrix is compressed into a 220×40 matrix and, as was explained, the acceleration response can be individually considered in the objective function as follows:

$$M_{i,j}\left(a_g\right) \xrightarrow{\text{Compressed}} N_{i,k}\left(a_g\right) = \begin{bmatrix} S^a_{1,1} - S^{aT}_{1,1} & \cdots & S^a_{1,l} - S^{aT}_{1,l} \\ \vdots & \ddots & \vdots \\ S^a_{m,1} - S^{aT}_{m,1} & \cdots & S^a_{m,l} - S^{aT}_{m,l} \end{bmatrix}$$

(4.12)

It should be noted that, if the size of the compressed matrix exceeds the limitations, the Out of Memory error will occur (Nozari and Estekanchi 2011). In order to compare the convergence of the current procedure (least squares optimization) with the previous procedure (unconstrained optimization), the optimization is performed utilizing identical initial points for both procedures, and the results of 200 iterations are considered. As was explained, the optimization is a time-consuming process, and 200 iterations take more than 120 hours in the old procedure. In Figure 4.5, the results of both procedures for the production of the first AF of the new series, are indicated. As can be seen, not only is the convergence rate of the new procedure about 10 times higher than the previous method, but also the accuracy of the new procedure is better.

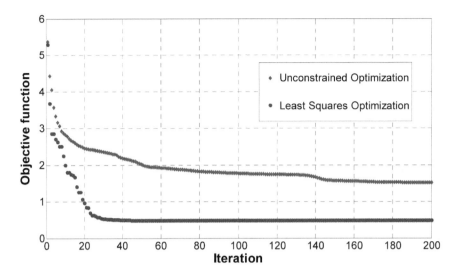

FIGURE 4.5 Comparison of convergence of least squares procedure (this work) with unconstrained procedure in optimization of ETA20d-TR01 acceleration function.

The produced ETEFs are named as ETA20d-TR01-03. The acceleration response spectra of the three ETEFs along with the average response spectra of the ETEFs, at the target time (the 10th second), are presented in Figure 4.6. A similar level of convergence exists approximately at all other times.

The absolute errors of responses of the new ETEFs, for both methods, the error in optimized points and the base error, are listed in Table 4.2. The absolute error of the average response of the three ETEFs is computed as well.

Comparing the results of the ETA20d and ETA20d-TR series of the ETEFs, it can be seen that the average error of the first series of ETEFs is 0.6284 m/s². This error for the second series ETEFs is 0.5009 m/s², that is, it is reduced by about 20 percent. In addition, the error of average response of the first series ETEFs is 0.4564 m/s². While this error for the new ETEFs is 0.3098 m/s², which is about 27 percent improved. It can be concluded that the utilization of the three ETEFs reduces the error by more than 38 percent as well.

4.5   OPTIMIZATION OF LONG DURATION ET ACCELERATION FUNCTIONS

The duration of the ETEFs produced previously was 20 seconds. If the duration of the ETEFs is increased from 20 seconds to 40 seconds, the number of acceleration points, as the optimization variables, is increased from 2,048 points to 4,096 points. Consequently, the efficiency of the optimization procedure seriously declines. On the other hand, by increasing the size of the objective matrix, the utilization of generated code for least squares optimization is impractical. Since, for transforming the initial objective matrix into a compressed matrix, a few objective times could be subjected to the optimization, which increases the errors and leads to degraded

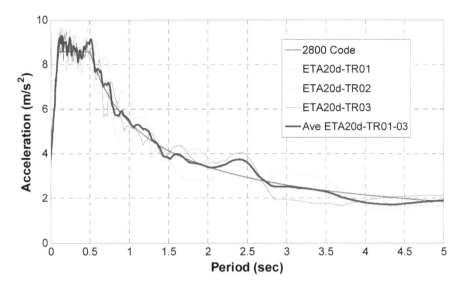

FIGURE 4.6 Acceleration response spectra of ETA20d-TR01-03 at the target time (the 10th second).

TABLE 4.2
Errors of Acceleration Responses of the Second Series ETEFs

Acceleration functions	Absolute error (m/s²)	
	Optimized points	**Base points**
ETA20d-TR01	0.5132	0.5095
ETA20d-TR02	0.4896	0.4993
ETA20d-TR03	0.4999	0.4941
Average	0.5008	0.5009
Ave ETA20d-TR01-03	0.3041	0.3098

results. Furthermore, due to the increase of number of variables, the computational demand is significantly increased; as a result, the convergence rate is decreased. Therefore, to produce ETEFs with longer duration, other techniques are required to be applied. Some mathematical transformation methods such as wavelet transforms (Mashayekhi, Estekanchi, and Vafai 2019a) or special base functions (Mashayekhi, Estekanchi, and Vafai 2019b) can be very useful for this purpose. However, in order to keep things simple, an idea used for this issue here is utilizing the same number of 2,048 acceleration points with 0.02-second time steps for production of 40-second ETEFs, and subsequently transforming them into the ETEFs with 4,096 acceleration points with 0.01-second time steps. After the primary experiments were conducted, a multilevel optimization procedure was adopted for production of the 40-second ETEFs as described in the following:

In the first step, a 20-second AF is produced by 2,048 acceleration points with a time step of 0.01 seconds. The process of optimization is similar to the process applied for the optimization of the second series of ETEFs. Subsequently, this AF is chosen as the initial point, and the optimization is performed by the time step of 0.02 seconds. Hence, a 40-second AF is produced with 2,048 acceleration points. In the next step, the number of acceleration points is increased to 4,096 acceleration points. For this purpose, an average of two acceleration points is considered as the acceleration numerical value for the time step between the two primary time steps. For example, the average of two acceleration values in 0.02 and 0.04 seconds is assumed as the 0.03 second acceleration value. Finally, 2,048 acceleration points resulted from the average of the primary optimized acceleration points and are considered as the variables of the optimization. Next, the objective function is assumed to keep the primary optimized points unchanged, and the optimization process is conducted on the acceleration points between them, 2,048 points, in a way that these points take place between the primary acceleration points. Afterwards, the objective function with 4,096 acceleration points and 0.01 seconds time step is calculated. The explained procedure is briefly illustrated in Figure 4.7.

In fact, by utilizing this procedure, a 40-second ETEF can be produced via two optimization cycles with 2,048 variables without dealing with the 4,096-variable optimization process. It should be noted that only the duration of these ETEFs has been enlarged with respect to 20-second ones, and both have the same response range; thus, the target time of a 40-second ETEFs is the 20th second, and their responses reach twice the codified values at the 40th second. The results of the optimization steps of the first 40-second ETAF by utilizing the aforementioned procedure are presented in Table 4.3. For each step, the numerical values of the base errors and the errors in optimized points have been calculated.

The improvement of the results can be observed in different steps of the procedure. It should be noted that, in the third step, the numerical value of base error is acceptable; however, the error for optimized periods has been increased. Due to the fact that these periods are chosen with the logarithmic distribution, and that the number of short periods is much more than long periods, and considering the sensitivity of the time-history analysis to the time steps in short periods, the amount of error is increased (Clough and Penzien 1993). Hence, the results in the third step are still not satisfactory, and performing an additional step of optimization is recommended in order to reduce errors to acceptable levels.

By applying this method, three 40-second ETEFs are produced and are called ETA40d01-03. A sample of these ETEFs and their acceleration and displacement response spectra at 10th, 20th, 30th, and 40th seconds are presented respectively in Figures 4.8 and 4.9. Moreover, the acceleration response spectra of the third series of the ETEFs and the average response spectrum of the three ETEFs, at the target time, (the 20th second) are depicted in Figure 4.10.

In Table 4.4, the absolute errors for the 40-second ETEFs are calculated considering two references: (1) at optimized points, (2) at base points. In addition, the absolute error for the average response of the three ETEFs is calculated. In Figure 4.11, the errors of the first, second, and third series of the ETEFs are compared, and the trend of error reduction from the first series to the third series could be clearly identified. In

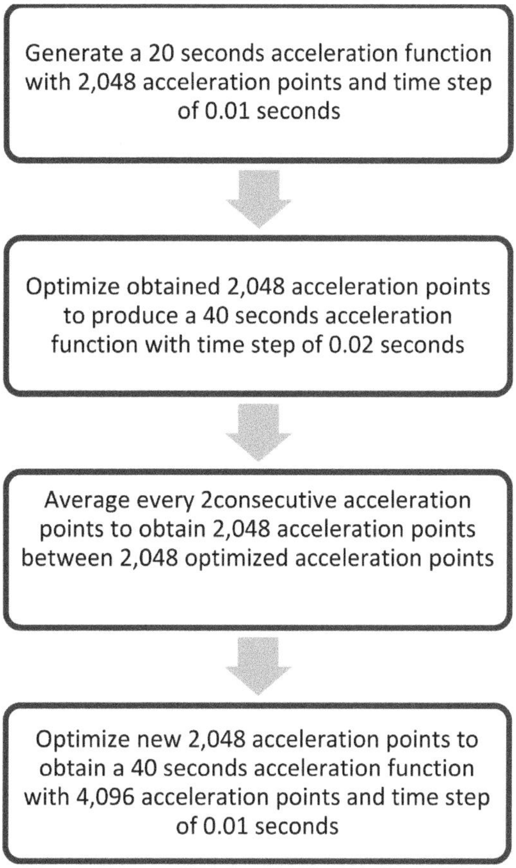

FIGURE 4.7   Optimization procedure of 40 seconds ET acceleration functions.

TABLE 4.3

Errors in Different Steps of Optimization of ETA40d01 Acceleration Function

Optimization step	Error in optimized points (m/s²)	Error in base points (m/s²)
1	0.5094	0.5378
2	0.3301	0.3647
3	0.5356	0.3741
4	0.3527	0.3705

this figure, the error reduction by utilizing the three ETEFs for each series is further evident.

In order to study the effect of expanding the duration of the ETEFs in their results accuracy, the errors of the ETA20d-TR and ETA40d series of ETEFs were compared. As can be observed, the average error of the 3-second series ETEFs is 0.5009 m/s²,

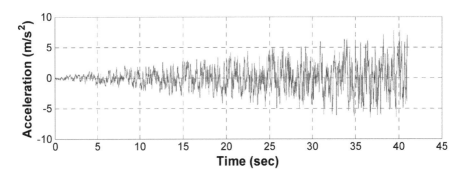

FIGURE 4.8 ETA40d01 acceleration function.

while the same numerical value for the third series ETEFs is 0.4006 m/s², which is reduced about 20 percent. Furthermore, the average response error of the 3-second series ETEFs is 0.3098 m/s², while the same numerical value for the third series ETEFs is 0.2263 m/s², which has improved by more than 26 percent.

By comparing the errors of the first and third series ETEFs, it can be observed that the average error of the first series is 0.6284 m/s², while the same numerical value for the third series is 0.4007 m/s², which is reduced by about 36 percent. In addition, the average response error of three first-series ETEFs is 0.4564 m/s², while the same numerical value for the third series ETEFs is 0.2263 m/s², which is improved by about 50 percent. It can further be observed that the utilization of three ETEFs reduces the error by more than 43 percent.

4.6 COMPARISON OF ETEFS IN THE ANALYSIS OF SDOF SYSTEMS

In this section, four SDOF systems with natural periods of 0.5, 1, 2, and 4 seconds are studied. The damping ratio of these systems is assumed to be 5 percent, in which the same value has been utilized to produce the ETEFs. These systems are analyzed with the ETEFs and the time-history responses are compared with the target time-history response calculated using the definition of the ET method based on the spectral response associated with the standard 2,800.

The ETA20d series of the ETEFs are applied, and the SDOF systems are analyzed, with the three ETEFs (ETA20d01-03). The acceleration and displacement responses of each system are calculated, and the average of the three responses is assumed as the final response of the system and is contrasted versus the target response. In Figure 4.12, the average acceleration and displacement responses of four SDOF systems are shown. Similarly, the SDOF systems are analyzed with the second series ETEFs (ETA20d-TR01-03) and the third series ETEFs (ETA40d01-03) and the results are compared with the target responses. Figures 4.13 and 4.14 indicate the average acceleration and displacement responses for the second and the third series ETEFs. As can be observed, the results obtained from the ET analysis with the second series of ETEFs are more consistent with the target response compared with the results of the first series ETEFs. Moreover, the consistency of results for the third

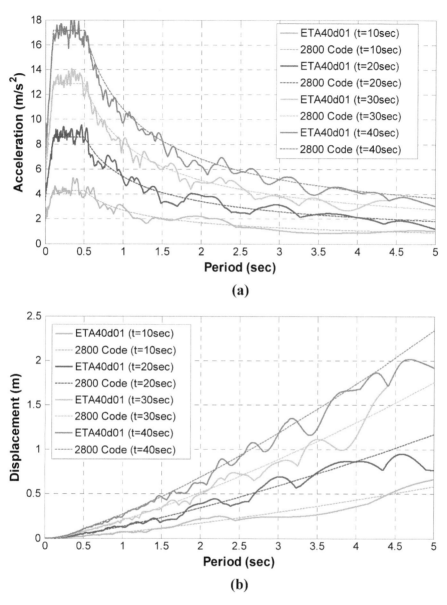

FIGURE 4.9 Response spectra of ETA40d01 at 10th, 20th, 30th, and 40th seconds, (a) acceleration, (b) displacement.

series ETEFs (40-second ETEFs), is more reasonable than the results obtained from both former series.

In Table 4.5, the acceleration responses errors of each SDOF system under the ETEFs are presented. Despite a number of exceptional cases, the descending trend of the error numerical values from the first to the third series of the ETEFs can be clearly

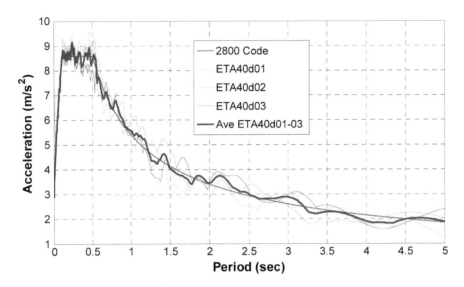

FIGURE 4.10 Acceleration response spectra of ETA40d01-03 at the target time (the 20th second).

TABLE 4.4
Errors of Acceleration Responses of the Third Series ETEFs

	Absolute error (m/s²)	
Acceleration function	**Optimized points**	**Base points**
ETA40d01	0.3527	0.3705
ETA40d02	0.3965	0.4142
ETA40d03	0.4193	0.4173
Average	0.3895	0.4007
Ave ETA40d01-03	0.2217	0.2263

identified. Specifically, the reduction of error for the third series ETEFs is more significant. For instance, as for the SDOF system with the period of 0.5 seconds, the average acceleration response error of the first and the second series ETEFs are close to each other, about 0.55 m/s², while the same numerical value for the third series ETEFs is about 0.30 m/s², which is improved by 45 percent.

In addition, for all the cases, the average response error of three ETEFs is reduced in comparison with the average of errors of three ETEFs responses. For example, as for the SDOF system with a 4-second period, the average error of responses resulted from the three first series ETEFs is about 0.60 m/s², while the error of average response of the first series ETEFs equals to 0.48 m/s², which indicates 20 percent of error reduction. This error reduction for the second and third series of the ETEFs is about 42 and 55 percent, respectively.

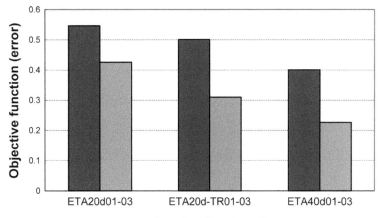

FIGURE 4.11 Comparison of three series of ET acceleration functions.

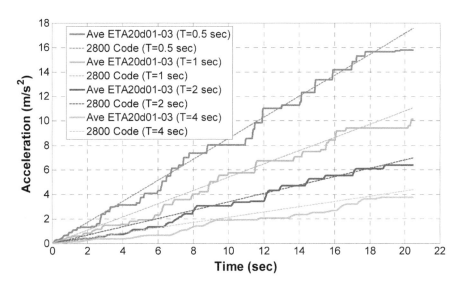

FIGURE 4.12 Acceleration response time – history of four SDOF systems for the first series ETEFs.

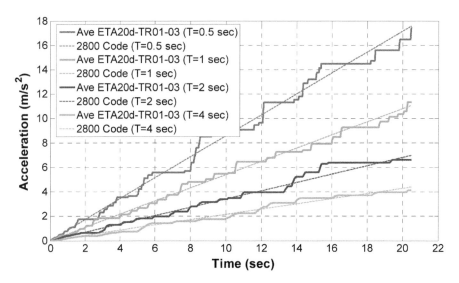

FIGURE 4.13 Acceleration response time history of four SDOF systems for the second series ETEFs.

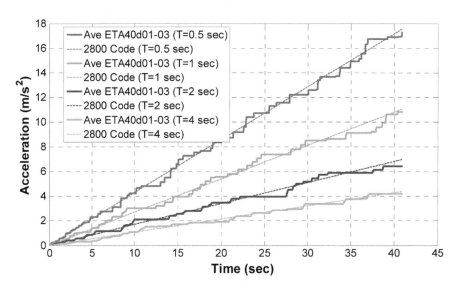

FIGURE 4.14 Acceleration response time history of four SDOF systems for the third series ETEFs.

TABLE 4.5

Acceleration Responses Errors for Four SDOF Systems under Three Series of the ETEFs

Acceleration function	Absolute errors of SDOF systems responses (m/s²)			
	T=0.5 sec	T=1 sec	T=2 sec	T=4 sec
ETA20d01	0.9076	0.6134	0.4235	0.4447
ETA20d02	0.8258	1.0034	0.6346	0.8646
ETA20d03	0.8993	0.7540	0.4803	0.5000
Average	0.8776	0.7903	0.5128	0.6031
Ave ETA20d01-03	0.5531	0.5278	0.3340	0.4783
ETA20d-TR01	1.0318	0.7001	0.3675	0.2767
ETA20d-TR02	0.7544	0.6256	0.5022	0.3362
ETA20d-TR03	0.7893	0.6298	0.4391	0.3773
Average	0.8585	0.6518	0.4363	0.3301
Ave ETA20d-TR01-03	0.5681	0.3033	0.3283	0.1887
ETA40d01	0.4913	0.4135	0.3633	0.2707
ETA40d02	0.6306	0.4664	0.3768	0.2986
ETA40d03	0.5520	0.7400	0.4076	0.2998
Average	0.5580	0.5400	0.3826	0.2897
Ave ETA40d01-03	0.2997	0.3222	0.2330	0.1226

4.7 SUMMARY AND CONCLUSIONS

Reliable simulation of seismic structural response is among the major challenges in earthquake engineering. The intensive computational demand is a considerable issue in practical applicability of many realistic simulation procedures capable of including complicated structural responses such as material and geometric nonlinearity. The ET method is a tool for the seismic design of structures in which they are subjected to a gradually intensifying dynamic excitation, and their seismic performance is evaluated based on its response at different excitation levels. Consequently, substantial reductions in computational demand can be achieved when structural performance at various excitation intensity levels is to be predicted. Generating appropriate artificial dynamic excitations is essential for the ET method's success. In this chapter, the basic numerical procedure for generating the ETEFs, and its formulation as a numerical optimization problem, was presented. The Trust Region algorithm utilized in the developed optimization program exhibits a high convergence rate in the optimization of the ETEFs. By the discrete time formulation of the ET method and defining the objective function of optimization in the matrix form, considering the computational limitations, the second series of ETEFs are produced in the linear range of structural analysis. It should be noted that the required time for the optimization of these ETEFs is nearly one-tenth of the time spent for the optimization of the original ETEFs, while, the average error of the second series of ETEFs is about 30 percent less than the average error of the first series of ETEFs.

Moreover, a procedure for extending the total duration of ETEFs without compromising accuracy and time efficiency was presented. The convergence and level of accuracy to be expected from generating the ETEFs was discussed by applying the generated ETEFs to SDOF systems as well. It can be concluded that the proposed procedures can be applied successfully to generate usable intensifying ETEFs, which should be applied in response-history analysis of structures utilizing the ET methodology. Generating optimal ETEFs is an open problem and can be approached using many different conceptual formulations and optimization procedures. This chapter serves as an introduction to understanding the basic and simple forms of formulation of the problem. Interested readers should follow the latest literature on the topic to become familiarized with the state of the art in this area.

4.8   NOMENCLATURE

a_g	Acceleration function
B	Building response factor
dt	Time step of time-history analysis
ET	Endurance Time method
$F(a_g)$	Objective function of optimization
g	Gravitational acceleration
I	Building importance factor
l	Number of discrete times
$M(a_g)$	Objective matrix of optimization
$N(a_g)$	Compressed objective matrix of optimization
m	Number of period points
n	Number of acceleration points
p	Integer parameter to determine time intervals
R	Response reduction factor
S_a	Spectral acceleration
S_{aC}	Codified acceleration response
S_{aT}	Target acceleration response
$S^a_{i,j}$	Maximum acceleration response for period T_i until time t_j
S_u	Spectral displacement
S_{uT}	Target displacement response
$S^u_{i,j}$	Maximum displacement response for period T_i until time t_j
T	Free vibration period
t	Time
t_{Target}	Target time
$u(t)$	Displacement response-time history
$\ddot{u}_g(t)$	Ground acceleration time history
α	Weighting parameter in objective function of optimization
ω_n	Natural circular frequency
ξ	Damping ratio

NOTE

1 *Chapter Source*: Nozari, A., and H.E. Estekanchi. 2011. "Optimization of Endurance Time Acceleration Functions for Seismic Assessment of Structures." *International Journal of Optimization in Civil Engineering* 1, no. 2, pp. 257–277.

REFERENCES

ASCE7-05. 2005. *Minimum Design Loads for Buildings and Other Structures*. Reston, VA: American Society of Civil Engineers.

BHRC. 2005. *Iranian Code of Practice for Seismic Resistant Design of Buildings*. Standard No. 2800-05, 3rd ed. Tehran: Building and Housing Research Center.

Clough, R.W., and J. Penzien. 1993. *Dynamics of Structures*. New York: McGraw-Hill.

Coleman, T.F., and Y. Li. 1994. "On the Convergence of Reflective Newton Methods for Large-scale Nonlinear Minimization Subject to Bounds." *Mathematical Programming* 67, no. 2, pp. 189–224.

Coleman, T.F., and Y. Li. 1996. "An Interior Trust Region Approach for Nonlinear Minimization Subject to Bounds." *SIAM Journal on Optimization* 6, no. 2, pp. 418–445.

Estekanchi, H.E. 2020. *Website of the Endurance Time Method*, [Online]. https://sites.google.com/site/etmethod/home.

Estekanchi, H.E., and M.C. Basim. 2011. "Optimal Damper Placement in Steel Frames by the Endurance Time Method." *The Structural Design of Tall and Special Buildings* 20, no. 5, pp. 612–630. Wiley Online Library. doi:10.1002/tal.689.

Estekanchi, H.E., V. Valamanesh, and A. Vafai. 2007. "Application of Endurance Time Method in Linear Seismic Analysis." *Engineering Structures* 29, no. 10, pp. 2551–2562. doi:10.1016/j.engstruct.2007.01.009.

Mashayekhi, M., H.E. Estekanchi, and H. Vafai. 2019(a). "Simulation of Endurance Time Excitations via Wavelet Transform." *Iranian Journal of Science and Technology*, Transactions of Civil Engineering, 43, no. 3, pp. 429–443.

Mashayekhi, M., H.E. Estekanchi, and H. Vafai. 2019(b). "Simulation of Endurance Time Excitations Using Increasing Sine Functions." *International Journal of Optimization in Civil Engineering* 9, no. 1, pp. 65–77.

Mashayekhi, M., H.E. Estekanchi, and A. Vafai. 2020. "Optimal Objective Function for Simulating Endurance Time Excitations." *Scientia Iranica*, 27, no. 4, pp. 1728–1739.

Mashayekhi, M., H.E. Estekanchi, H. Vafai, and G. Ahmadi. 2019. "An Evolutionary Optimization-based Approach for Simulation of Endurance Time Load Functions." *Engineering Optimization* 51, no. 12, pp. 2069–2088.

Nozari, A., and H.E. Estekanchi. 2011. "Optimization of Endurance Time Intensifying Acceleration Functions Using Trust Region Algorithm." Sixth National Conference on Civil Engineering. Semnan, Iran.

Riahi, H.T., and H.E. Estekanchi. 2010. "Seismic Assessment of Steel Frames with Endurance Time Method." *Journal of Constructional Steel Research* 66, no. 6, pp. 780–792. doi:10.1016/j.jcsr.2009.12.001.

Valamanesh, V., H.E. Estekanchi, and A. Vafai. 2010. "Characteristics of Second Generation Endurance Time Acceleration Functions." *Scientia Iranica* 17, no 1, pp. 53–61.

5 Correlating Analysis Time with Intensity Indicators

5.1 INTRODUCTION

Estimating the seismic performance of structures during and after earthquakes is always an important issue in the design of seismic-resistant structures.[1] This has motivated structural engineers to develop design concepts that permit the designer to investigate whether a design is capable of meeting the desired performance objectives during and after a specified earthquake. This approach was proposed by the SEAOC Vision 2000 Committee in 1995 (1995), and is called "Performance-Based Earthquake Engineering." Performance-based earthquake engineering is a methodology, in which structural design criteria are expressed in terms of achieving a set of different performance objectives (Ghobarah 2001). Performance objective is a practical notion that consists of the specification of a structural performance level (e.g., collapse prevention (CP), life safety (LS), or immediate occupancy (IO)) for a given level of seismic hazard. For example, in accordance with SAC 2000 (2000) ordinary buildings are expected to provide less than a 2-percent chance over 50 years of damage of exceeding CP performance (Krawinkler and Miranda 2004).

In the Endurance Time (ET) method, buildings are rated according to the length of time they can endure a standard, calibrated, intensifying accelerogram. Higher endurance time is associated to a more suitable seismic performance (Estekanchi, Vafai, and Sadeghazar 2004; Estekanchi et al. 2011). The application of the ET method in performance-based design was studied by Mirzaee, Estekanchi, and Vafai (2010). They proposed a continuous performance curve called the "Target Performance Curve," which expresses the intended limit of the seismic performance of a structure at different seismic intensities. By comparing the performance curve of a structure obtained from ET analysis with the target performance curve, the seismic performance of the structure at different seismic intensities can be evaluated.

Target performance curve and seismic performance curve from ET analysis can be presented as a function of analysis time. However, analysis time in ET analysis is representative of the applied seismic intensity. Therefore, substituting a commonly used seismic intensity parameter (such as PGA, return period, or annual probability

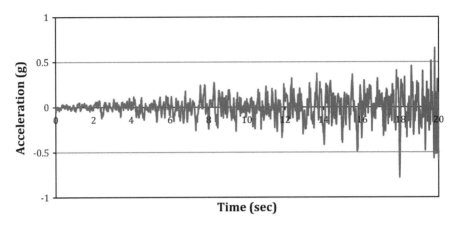

FIGURE 5.1   A typical ET accelerogram.

of exceedance) for time in the presentation of the seismic performance or other analysis results can make it more useful and convenient. This can also improve the readability of the seismic performance and target performance curves. To accomplish this goal, a correlation between time in ET analysis and the desired seismic intensity parameter should be established. In this chapter, the correlation between time in ET analysis and the seismic intensity return period, which is an important term in the definition of performance objectives, will be investigated. The ETA20jn series of ET accelerograms will be used for the purpose of demonstration of the concept. This endurance time excitation function (ETEF) represents a typical code design spectra-compliant ETEF. It is created in such a way that its response spectrum at $t=10$ sec matches to the ASCE41 design spectrum matched to a typical site in Tehran (Mirzaee and Estekanchi 2011).

5.2   ENDURANCE TIME METHOD

ET accelerograms are produced in such a way that the amplitude of the acceleration is increased over time (Figure 5.1). Hence, in this method, each value of the time variable is representative of a particular seismic intensity. The ET accelerogram is conventionally created such that at a predefined time, t_{Target}, its response spectrum reaches a prespecified template response spectrum with a scale factor of unity. For example, three accelerograms, named "ETA20jn01-03," are created in such a way that their response spectrum at $t=10$ sec is compatible with the ASCE41 design spectrum adapted for a Tehran site (Mirzaee and Estekanchi 2011). The duration of these series of ET accelerograms is 20 seconds. They can be applied for nonlinear analysis due to the fact that their spectrum is matched to a target spectrum in the long period range. In this section, the parameter of time in ET accelerogram will be correlated to its equivalent intensity level.

ET accelerograms can have different intensification schemes. A linear intensification scheme has been conventionally applied for producing ET excitations. In this model, the response spectrum of an ET accelerogram is to intensify proportionally with time. Consequently, the target acceleration response of an ET accelerogram is defined as in Equation 5.1.

$$S_{aT}(T,t) \equiv S_{aC}(T) \times \frac{t}{t_{Target}} \tag{5.1}$$

Here $S_{aT}(T,t)$ is the target acceleration response at time t, T is the period of free vibration, and $S_{aC}(T)$ is the codified design acceleration spectrum (Estekanchi, Valamanesh, and Vafai 2007). This formula simply illustrates the linear proportionality between acceleration response produced by ET accelerogram at a particular time t and the considered template spectrum. It should be considered that a linear intensification with time is only being assumed in order to synthesize an elementary ET acceleration function (Estekanchi et al. 2011). Other intensification profiles can also be used based on the context of the problem (Mashayekhi et. al. 2018).

The results of the ET method are usually interpreted by a curve called "ET Response Curve" or "ET Performance Curve." Figure 5.2 shows the ET response curves for two steel moment frames. In this figure, the maximum interstory drift ratio is utilized as a representative of the performance of the frames. As will be shown, by correlating the analysis time to a convenient intensity parameter and using a conventional indicator of intensity in the horizontal axis, a much more convenient and self-explanatory graph can be obtained.

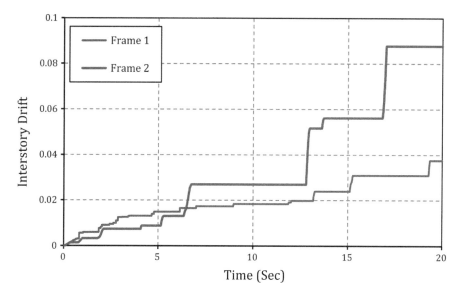

FIGURE 5.2 ET performance curves for two steel moment frames.

5.3 MULTIPLE INTENSITY LEVELS IN THE PERFORMANCE-BASED DESIGN

Performance-based seismic engineering (PBSE) is a methodology by which structural design criteria are expressed in terms of achieving a set of different performance objectives at different seismic intensity levels (SEAOC Vision 2000). In fact, the promise of PBSE is to produce structures with predictable seismic performances (Naeim, Bhatia, and Lobo 2001). Performance objective is a practical notion, which consists of the specification of a structural performance level (e.g., CP, LS, or IO) for a given level of seismic hazard. For example, according to SAC 2000, ordinary buildings are expected to provide less than a 2-percent chance over 50 years of damage exceeding CP performance (Krawinkler and Miranda 2004). A performance objective can include determination of different performance levels for several levels of seismic hazard. This type of performance objective is called "dual" or "multiple-level." A key parameter in determination of the performance objective is the return period, since it is proportional to the intensity level of expected earthquake. The return period is defined as the average period of time, in years, between the expected occurrences of an earthquake of specified intensity.

Application of the ET method in performance-based design has been studied by Mirzaee et al. (2010). Since ET analysis produces dynamic response as a continuous function of time, the authors introduced a curve called the "Target Curve," which expresses the limit of the desired seismic performance of a structure (acceptance criteria) at various times in ET analysis (note that times in ET analysis can be interpreted as seismic intensities). By comparing the ET performance curve with the target curve, the seismic performance of the structure at different seismic intensities can be evaluated (See Figure 5.3). The target curve

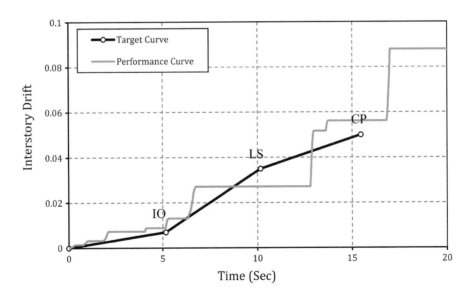

FIGURE 5.3   An example of target and performance curves [6].

TABLE 5.1
Equivalent Time Corresponding to Each Performance Level

Performance level	Probability of exceedance	Mean return periods (years)	PGA(g)	Endurance time (sec)	Interstory drift (%)
IO	50%/50 years	75	0.22	5.16	0.7
LS	10%/50 years	475	0.35	10.16	3.5
CP	2%/50 years	2,475	0.53	15.46	5

was created by linking each performance level (IO, LS, and CP) to ET analysis time using PGA (Peak Ground Acceleration) as an intermittent parameter. Other intensity indicators such as Sa can also be used in a similar way. This procedure is inclusive of three following steps:

1. Using an appropriate Gutenberg-Richter equation to obtain the magnitude corresponding to mean return period related to each of the three performance levels.
2. Acquiring the peak ground accelerations for the considered site based on an attenuation relationship utilizing the previously identified magnitudes.
3. Identifying the equivalent ETs in ET records, corresponding to the three mentioned PGA (This means tracing the times in the ET acceleration function at which the PGAs exceed the values of the PGAs corresponding to each performance level).

Typical performance objectives and the analysis times correlated to each seismic intensity are shown in Table 5.1. In this table, the relevant maximum interstory drift for each performance level is also indicated (Mirzaee, Estekanchi, and Vafa 2010). These quantities are typical values for steel moment frames based on the FEMA-356 Prestandard and Commentary for the Seismic Rehabilitation of Buildings (FEMA-356 2000).

5.4 CORRELATION BETWEEN TIME IN ET ANALYSIS AND RETURN PERIOD

Both the performance and target curves represent the variation in the seismic performance of the structures with time. This means that by utilizing such curves, the seismic performance can be evaluated at specific times in ET analysis correlated to specific seismic intensity matching the desired performance objective. While time is an indicator of intensity in ET analysis, the corresponding intensity depends on the specifications of the ET acceleration function used, as well as some other factors. Therefore, ET analysis time is not an obvious intensity measure, and better indicators of intensity can be used to define the variation of seismic performance with time. Therefore, substituting a common intensity parameter (such as PGA, return period, or annual probability of exceedance) for ET analysis time in the evaluation and expression of the performance of the structures is highly desirable. By this sub-stitution, the performance and target curves will be more explicit, and their inter-pretation will be more obvious. To make such a substitution, a correlation between ET analysis time and the stated common parameter should be determined. In this section, the problem of correlation between the time and the return period of ground motion will be explained. The ETA20jn series of ET accelerogram is used as the basic accelerogram in this investigation. It should be noted that normally, this correl-ation depends on the fundamental period of the structure. In other words, different structures with different fundamental periods will have different ET analysis times, relevant to a particular return period (Bazmooneh 2009).

To establish such an interrelationship, the response spectrum will be utilized as an intermittent criterion in this section. The time at which the response spectrum is matched to the response spectrum corresponding to a particular hazard level (or return period) is traced. This procedure can be accomplished by various approaches. One approach is to match the two aforementioned response spectra at the funda-mental period of the structure. Another approach is to use a range of periods, instead of the fundamental structure period (Bazmooneh 2009). In this chapter, both methods will be considered and compared with each other.

The response spectrum for a particular hazard level and a specific site can be obtained using a standard building code. Herein, the ASCE Standard for Seismic Rehabilitation of Existing Buildings, known as ASCE-41 (Bazmooneh 2009, American Society of Civil Engineers 2006), is considered, and the spectra are acquired for Tehran. It should be pointed out that the site is classified as Site Class C with $V_{s30} \approx 600$ m/s and is generally similar to typical sites in the Los Angeles area.

In the definition of the ASCE-41 Standard design response spectrum, two spec-tral response acceleration parameters should be clearly determined, namely, S_s (short-period spectral response acceleration parameter) and S_1 (long-period spectral response acceleration parameter). These two parameters are usually obtained by util-izing Maximum Considered Earthquake (MCE) maps. Since these maps are not avail-able for Tehran at present, site-specific procedures should be used to obtain these spectral parameters. In this research, the seismic hazard curves for S_s and S_1 obtained by Mirzaee and Estekanchi (2011) are used to develop a formulation for S_a (spectral acceleration) versus return period; subsequently, the response spectra for different hazard levels are drawn.

In Figure 5.4, the assumed seismic hazard curves for S_s and S_1, for Tehran, are shown. According to this figure, the relation between S_s and S_1 and the annual prob-ability of exceedance (λ_m) can be derived, as in Equations 5.2 and 5.3.

$$S_s = 0.072 \times \lambda_m^{-0.43} - 0.2674 \qquad (5.2)$$

$$S_1 = 0.026 \times \lambda_m^{-0.44} - 0.16 \qquad (5.3)$$

Considering that the annual probability of exceedance is equal to the inverse of the return period (Kramer 1996), the ASCE41 response spectrum can be introduced, according to the value of the return period, as indicated in Equation 5.4.

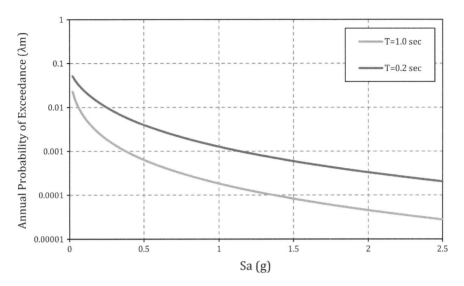

FIGURE 5.4   Seismic hazard curve for S_a

$$S_a = f(R,T) = \begin{cases} \dfrac{0.45T(R^{0.43}-3.7)^2}{R^{0.44}-6.17} + 0.029(R^{0.43}-3.74) & T < T_0 \\ 0.072R^{0.43}-0.27 & T_0 < T < T_S \\ \dfrac{0.034R^{0.44}-0.21}{T} & T_S < T \end{cases} \qquad (5.4)$$

where,

S_a = spectral acceleration
T = period of free vibration
R = return period

$$T_S = \frac{R^{0.44}-6.17}{2.12R^{0.43}-7.90}$$

$$T_0 = 0.2T_S = \frac{R^{0.44}-6.17}{10.6R^{0.43}-39.51}$$

Therefore, the ASCE41 response spectrum for any hazard level can be obtained, based on the above equation. In Figure 5.5, the response spectra at various hazard levels are illustrated.

Acquiring the inverse of function $f(R, T)$ given in Equation 5.4, with respect to variable R, the return period can be expressed as a function of T and S_a (Equation 5.5).

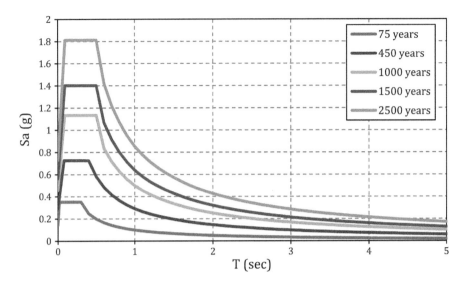

FIGURE 5.5 The ASCE41 response spectra for Tehran, for various hazard levels.

$$R = h\left(S_a, T\right) = f^{-R}\left(S_a, T\right) \tag{5.5}$$

Here R is the return period, $h(S_a, T)$ is a function that relates the return period to S_a and T, and f^{-R} represents the inverse of function f (given in Equation 5.4), with respect to variable R.

Thus, the values of the return period can be derived from the values of T and S_a. Since establishment of an explicit formulation for this function is not straightforward, a matrix has been developed for the return periods, where, at each structural period and each S_a, the matching return period is specified.

The response spectrum for the ET accelerogram is a function of time and is define as indicated in Equation 5.6 (Estekanchi, Valamanesh, and Vafai 2007):

$$S_a\left(T, t\right) = \max\left(\left|a\left(\tau\right)\right|\right) \qquad\qquad \tau \in \left[0, t\right] \tag{5.6}$$

where T is the period of free vibration, t is time, and a is acceleration.

As can be seen from Equations 5.5 and 5.6, S_a is dependent on T and t, and the return period has been developed as a function of T and S_a. Therefore, the return period can be expressed as a function of T and t, accordingly. As mentioned before, since expressing the return period via an explicit formulation is a complex process, this function can be numerically represented by a matrix called A_{RP}, as shown in Equation 5.7. To develop the matrix A_{RP}, the value of S_a is calculated for the intended T and t using Equation 5.6 (or Equation 5.1). Then the desired return period can be calculated utilizing Equation 5.5 regarding the values of intended T and obtained S_a.

$$
A_{RP} = \begin{array}{c}
\\ \\ 5 \\ \\ \\ 10 \\ \\ \\ 17 \\ \\ \\ \\
\end{array}
\overset{\begin{array}{ccccccc} & & 1 & & 3 & & 5 \end{array}}{\begin{bmatrix}
\cdots & & \cdots & & \cdots & & \cdots \\
\cdot & & \cdot & & \cdot & & \cdot \\
\cdot & & \cdot & & \cdot & & \cdot \\
\cdot & & \cdot & & \cdot & & \cdot \\
\cdots & 220 & \cdots & 187 & \cdots & 83 & \cdots \\
\cdot & & \cdot & & \cdot & & \cdot \\
\cdot & & \cdot & & \cdot & & \cdot \\
\cdot & & \cdot & & \cdot & & \cdot \\
\cdots & 490 & \cdots & 422 & \cdots & 408 & \cdots \\
\cdot & & \cdot & & \cdot & & \cdot \\
\cdot & & \cdot & & \cdot & & \cdot \\
\cdots & 1108 & \cdots & 878 & \cdots & 868 & \cdots \\
\cdot & & \cdot & & \cdot & & \cdot \\
\cdot & & \cdot & & \cdot & & \cdot \\
\cdot & & \cdot & & \cdot & & \cdot \\
\end{bmatrix}} \qquad (5.7)
$$

Along the left axis: ET analysis time (sec). Along the top: T (sec).

Based on the above calculations, the return period can be obtained by either of two techniques: (1) T_0 method, in which, for each structure with a specific period of free vibration, the return period is computed at the fundamental period of the structure. (2) VT_0 method, in which, for each structure with a specific period of vibration, the return period is computed as an average over a vicinity of the fundamental period (0.2 to 1.5 times the fundamental period is considered). Equivalent return periods can be expressed by Equations 5.8 and 5.9 in these cases, respectively:

$$
R_{T_0}(t) = \left\{ r \mid S_a(T_0, t) = f(r, T_0) \right\} \qquad (5.8)
$$

$$
R_{VT_0}(t) = \left\{ r \mid \int_{0.2T_0}^{1.5T_0} S_a(T, t)\, dT = \int_{0.2T_0}^{1.5T_0} f(r, T)\, dT \right\} \qquad (5.9)
$$

Where $R_{T_0}(t)$ and $R_{VT_0}(t)$ are the return periods corresponding to time t in each method, and other parameters are as defined before. Figure 5.6 graphically shows the difference between the results of these two methods, where the response spectrum for the return period of a thousand years is depicted and fitted to the ET response spectra for a structure with the fundamental period of T_0. As illustrated in this figure, if the T_0 method is used, the corresponding time in ET analysis to the return period of a thousand years will be 15 seconds (the intersection is shown by a red cross), whereas the VT_0 method leads to an ET analysis time equal to about 14 seconds. Herein, these two

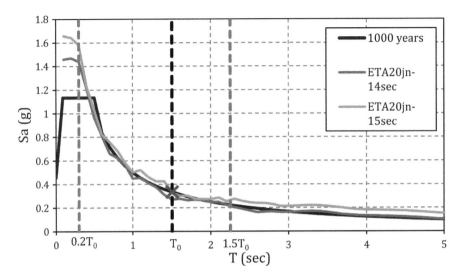

FIGURE 5.6   Correlating between seismic hazard and ET response spectra using T_0 and VT_0 methods.

methods are employed to relate the time in ET analysis to the return period, and the results are compared.

Figure 5.7 shows the return period versus time for various values of T, using the two aforementioned methods. As illustrated in this figure, for the structural periods greater than about two seconds, the curves are so close that a single curve can be used instead. That is, for the structures with periods greater than two seconds, the effect of the fundamental period of the structure can be eliminated. This situation is generally true for periods less than 0.2 seconds. The reason for this phenomenon can be implied by examining Figure 5.7, where the response spectra are very close to each other for periods less than 0.2 seconds and greater than two seconds. Moreover, as expected, the return period is increased as ET analysis time increases.

Considering Figure 5.7, it can be observed that the significance of the structural period increases as ET analysis time is increased – that is, as the seismic intensity is increased. In addition, for a particular return period, time in ET analysis increases as the period of the structure increases. It means that, as the period of the structure is increased, the structure should be subjected to the ET accelerogram for longer time, in order to experience a shaking equivalent to an earthquake with a specified return period. This is more obvious when the average method is applied. The reason behind these observations lies behind the differences in the considered hazard curves and the template spectra used in producing the ET acceleration functions.

Figure 5.8 shows the variation of the return period with the period of the structure for various ET analysis times. As can be seen, in this figure, the return period is decreased as the period of the structure is increased. Since the second method yields smoother curves, it is more appropriate to utilize it.

Figure 5.9 illustrates the variation of the return period with the structural period and time in ET analysis. In this figure, increases in the return period due to increases

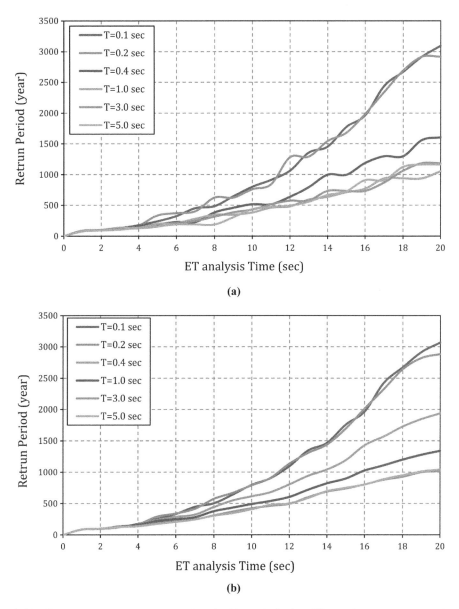

FIGURE 5.7 Return period versus time in ET analysis for different fundamental periods (a) T_0 method (b) VT_0 method.

in the structural period and ET analysis time are shown simultaneously. As depicted in this figure, the maximum return periods are observed at lower structural periods and, naturally, at higher times.

It should be noted that the effect of the fundamental period of the structure on the relation between ET analysis time and return period is strongly dependent on the type

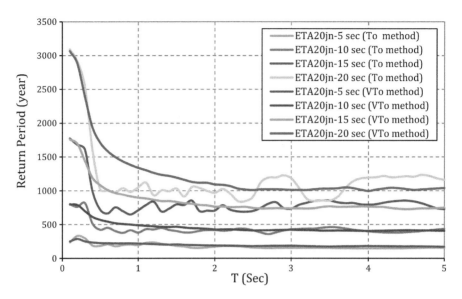

FIGURE 5.8   Return period versus fundamental period for various ET analysis times (T_0 and VT_0 methods).

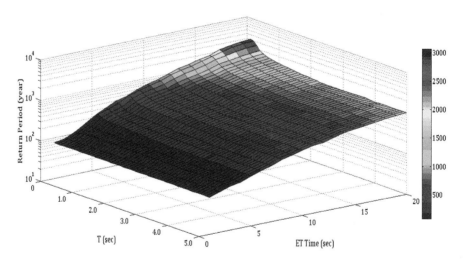

FIGURE 5.9   Return period versus structural period and ET analysis time (Average method).

of ET accelerogram. Ideally, and theoretically, if an ET accelerogram is generated in such a way that its response spectra are at any time completely coincident with a specific seismic hazard response spectrum, the dependency of return period to the structural period will be dropped.

The ET accelerogram used here has been produced so that its response spectrum has higher values at lower periods compared to the response spectrum of seismic

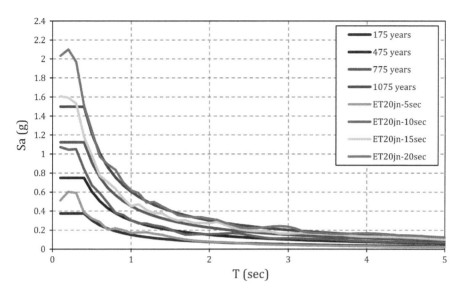

FIGURE 5.10 ETA20jn and hazard levels' response spectra.

hazard levels (See Figure 5.10). Thus, for a specific time in ET analysis with a particular response spectrum, the correlated return period for lower structural periods is
significantly greater than for higher structural periods. This result can also be inferred
from Figures 5.8 and 5.9.

This issue is important and can play a vital role in the development of new
generations of ET accelerograms. Producing an accelerogram, in such a way that its
response spectrum for some ET analysis times becomes compatible with the design
response spectra of significant hazard levels at all periods, will highly improve versatility of such accelerograms in the seismic assessment of structures, which is probably
feasible.

5.5 EXPLANATORY CASE STUDY

The methodology introduced in this chapter will be explained by considering a typical three-story steel moment-resisting frame designed according to AISC-ASD
building code (American Institute of Steel Construction 1989). This frame has been
intentionally designed as weak by considering one-half of the design base shear
recommended by ASCE41 (considering a design spectrum with S_s, S_1, F_a and F_V equal
to 0.768, 0.229, 1.093, and 1.57, respectively), and will be hereafter referred to as
the MF3S1. The frame will be studied by subjecting it to "ETA20jn" series of ET
accelerograms. The basic properties of this frame are provided in Table 5.2. As previously mentioned, the "ETA20jn" of ET accelerograms is created in such a way that
the response spectrum at t=10 sec is compatible with the ASCE41 (American Society
of Civil Engineers 2006) design spectrum that roughly corresponds to a typical site in
Tehran with soil condition considered as Site Class C.

TABLE 5.2
MF3S1 Frame Basic Properties

Property	Value
Number of stories	3
Number of spans	1
Mass participation (Mode 1)	84.7%
Period of free vibration (sec)	2.075
Design base shear over the weight	0.0675
Column sections	HE140B
Beam sections	HE160A

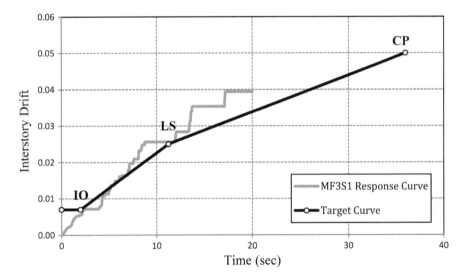

FIGURE 5.11   Performance curve for MF3S1 by ET analysis time.

The modeling and nonlinear analysis were performed using PEER's OpenSees platform (Pacific Earthquake Engineering Research Center 2004). The ET performance curve of the frame is obtained and its performance is evaluated by comparing this curve with the former target curve, which illustrates the acceptance criteria versus time. The result is shown in Figure 5.11. By utilizing the transformation obtained in Section 5.4 for return periods versus ET analysis times, the return period has been substituted for time in Figure 5.12, and the new target and performance curves are depicted.

Comparison of the new target and performance curves with former ones reveals that while they basically show the same information, the seismic performance of the structures has been more clearly interpreted in Figure 5.12, since it has replaced a redundant time axis with a more relevant return period axis. The new curves are more understandable and more useful, since the damage index (here: interstory drift) can be directly obtained for each return period of interest. For this frame, the value

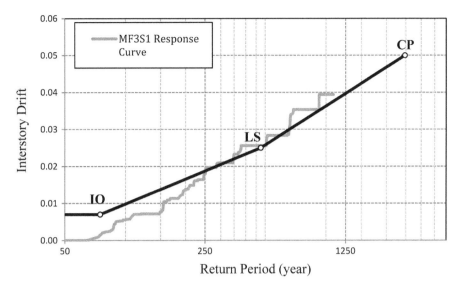

FIGURE 5.12 Performance curve for MF3S1 by return period.

of interstory drift at low hazard levels is acceptable but, for the hazard levels higher than 200 years, this parameter is somewhat above the target performance criteria and one can say that the frame does not perform well in this area. In other words, for the earthquakes of return period greater than two hundred years, this design is not acceptable according to the specified criteria. Although at a return period of about 500 years the frame seismic performance is almost satisfactory.

In order to explain the versatility of proposed methodology, the application of viscous dampers is considered in order to improve performance. A viscous damper is placed at the ground floor, and the new frame (referred to as MF3S2 hereafter) is subjected to the same series of ET accelerograms. To model the viscous damper, viscous material available in OPENSEES is used. The induced stress in this material is acquired from the following equation:

$$\sigma = C_0 \, |\dot{\varepsilon}|^\alpha \, sign\left(\dot{\varepsilon}\right) \tag{5.10}$$

where σ represents the induced stress in the material, $\dot{\varepsilon}$ is strain rate, C_0 is damping ratio, and α is damping exponent. In MF3S2, C_0 equals 100MPa and α equals 1. The seismic performance of two frames has been compared by ET response curves shown in Figure 5.13. As the figure shows, the frame performance has been improved within a wide range of seismic intensities, which is regarded as the consequence of utilizing a viscous damper. This form of presentation of ET analysis results is more suitable for practical applications by the structural designer. The target curve represents which drift levels are to be considered as acceptable by performance criteria. While resultant drifts need to be compared to allowable ones only at specific points marked as IO,

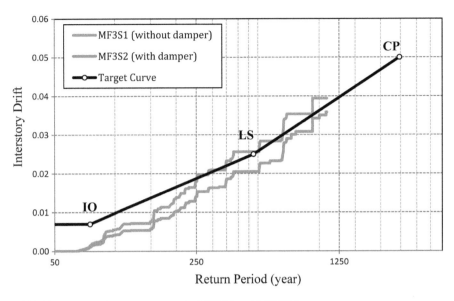

FIGURE 5.13   Performance curves for MF3S1 and MF3S2 frames.

LS, and CP considering code requirements; a continuous target performance curve conveys a better presentation of a desirable performance objective in general. Also, note that the comparison of relative performance of two different designs becomes much more intuitive using such diagrams.

In this case it should be noted that, as Figure 5.13 shows, the performance curve is obtained up to a return period equal to about 1,100 years, which is less than the return period of 2,475 years normally considered for the CP performance level. This issue points to the need to generate new ET accelerograms with longer durations to cover higher ground-motion intensities. The same ET accelerograms could also be upscaled to cover the desired intensity range; however, the accuracy of the analysis in nonlinear range might be compromised and is not considered in this explanatory example.

5.6   SUMMARY AND CONCLUSIONS

In this chapter the correlation between ET analysis time and the return period as an indicator of seismic intensity and/or risk was explained. The proposed procedure is based on the coincidence of response spectra obtained from ET accelerograms at different times and response spectra defined by ASCE41 at different hazard levels. It is more appropriate to compute the return period related to each time in ET analysis as an average over a range of structural periods, rather than computing at the fundamental period of the structure.

Results of the study suggest the following conclusions:

1. Substitution of the return period for time in the target and ET performance curves increases the readability of ET analysis results and can considerably

improve the presentation of ET analysis results in performance-based designs where responses at multiple seismic levels are to be studied.

2. The effect of the fundamental period of the structure on the relation between time in ET analysis time and the return period is strongly dependent on the compatibility of ET accelerogram template spectrum with design spectra at various intensity levels. Ideally, if an ET accelerogram is generated in such a way that its response spectra coincide with design seismic hazard response spectra, then the equivalent return period will only become a function of time instead of a function of time and fundamental period.

3. Due to the significant difference between the template spectrum of the ETA20jn series of ET accelerograms and the developed response spectra applying the ASCE41 approach for a typical Tehran site, the correlation between ET analysis times and the return period becomes dependent on the fundamental period of the structure. Although for periods less than 0.2 seconds and greater than two seconds, the effect of the period of the structure will be decreased, it is noticeable that this effect increases as time is increased, that is, as the seismic intensity is increased.

4. For this series of ET accelerogram, at a particular time the return period decreases with an increase in the period of the structure. Likewise, variation of ET analysis time is proportional to the period of the structure for a specific return period. Both these results mean that, as the period of the structure is increased, the structure should be subjected to an ET accelerogram for longer duration in order to experience shaking equivalent to an earthquake with a specified return period.

5. Generating ET accelerograms in which response spectrum at different ET analysis times remains compatible with the design response spectra would improve the application versatility of such accelerograms in seismic assessment.

5.7 NOMENCLATURE

$a_g(t)$	Ground acceleration
A_{RP}	Matrix of return period
CP	Collapse prevention
ET	Endurance time method
F_a	Site coefficient for S_S
F_V	Site coefficient for S_1
IO	Immediate occupancy
LS	Life safety
MCE	Maximum considered earthquake
PBSE	Performance-based seismic engineering
PGA	Peak ground acceleration
R	Return period
$R_{T_0}(t)$	Equivalent return period at time t calculated by T_0 method

$R_{VT_0}(t)$ Equivalent return period at time t calculated by VT_0 method
S_1 Long-period spectral response acceleration parameter
S_a Spectral acceleration
$S_{aC}(T)$ Code acceleration response for period T
$S_{aT}(T,t)$ Target acceleration response for period T at time t
S_s Short-period spectral response acceleration parameter
T   Free vibration period
t   Time
t_{Target} Target time
λ_m Annual probability of exceedance
σ Stress in damper
$\dot{\varepsilon}$ Strain rate
C_0 Damping ratio
α Damping exponent

NOTE

1 *Chapter Source*: Mirzaee, A., H.E. Estekanchi, and A. Vafai. 2012. "Improved Methodology for Endurance Time Analysis: From Time to Seismic Hazard Return Period." *Scientia Iranica* 19, no. 5, pp. 1180–1187.

REFERENCES

American Institute of Steel Construction (AISC). 1989. *Manual of Steel Construction: Allowable Stress Design*, 9th ed. Chicago.

American Society of Civil Engineers. 2006. *Seismic Rehabilitation of Existing Buildings*. SEI/ASCE 41-06, Reston, VA.

Bazmooneh, A. 2009. "Application of Endurance Time Method in Seismic Evaluation of Existing Steel Buildings." MS Thesis, Sharif University of Technology.

Estekanchi, H.E., A. Vafai, and M. Sadeghazar. 2004. "Endurance Time Method for Seismic Analysis and Design of Structures." *Scientia Iranica* 11, no. 4, pp. 361–370.

Estekanchi, H.E., A. Vafai, V. Valamanesh, A. Mirzaee, A. Nozari, and A. Bazmuneh. 2011. "Recent Advances in Seismic Assessment of Structures by Endurance Time Method." *Proceeding of a U.S-Iran-Turkey Seismic Workshop – Seismic Risk Management in Urban Areas*, pp. 289–301. PEER report, Istanbul.

Estekanchi, H.E., V. Valamanesh, and A. Vafai. 2007. "Application of Endurance Time Method in Linear Seismic Analysis." *Engineering Structures* 29, no. 10, pp. 2551–2562.

FEMA-356. 2000. *Prestandard and Commentary for the Seismic Rehabilitation of Buildings*. Washington, DC: Federal Emergency Management Agency.

Ghobarah, A. 2001. "Performance-Based Design in Earthquake Engineering: State of Development." *Engineering Structures* 23, pp. 878–884.

Kramer, S.L. 1996. *Geotechnical Earthquake Engineering.*, Upper Saddle River, NJ: Prentice Hall.

Krawinkler, H., and E. Miranda. 2004. "Performance-Based Earthquake Engineering." In *Earthquake Engineering: From Engineering Seismology to Performance-Based Engineering*, eds. Y. Bozorgnia and V.V. Bertero, Chapter 9, pp. 1–59. Boca Raton: CRC Press.

Mashayekhi, M., H.E. Estekanchi, A. Vafai, and S.A. Mirfarhadi. 2018. "Simulation of Cumulative Absolute Velocity Consistent Endurance Time Excitations." *Journal of Earthquake Engineering*, 25 no. 5, pp. 892–917.

Mirzaee, A., and H.E. Estekanchi. 2011. "Development of a Multilevel Design Response Spectrum for Tehran Using ASCE-41 Methodology and Logic Tree Approach." Submitted.

Mirzaee, A., H.E. Estekanchi, and A. Vafai. 2010. "Application of Endurance Time Method in Performance-Based Design of Steel Moment Frames." *Scientia Iranica* 17, no. 6, pp. 361–370.

Naeim, F., H. Bhatia, and R.M. Lobo. 2001. "Performance Based Seismic Engineering." In *The Seismic Design Handbook,* ed. F. Naeim. Springer, Boston, MA.

Pacific Earthquake Engineering Research Center (PEERC). 2004. *Open System for Earthquake Engineering Simulation* (OpenSees). Berkeley, CA: Pacific Earthquake Engineering Research Center (http://opensees.berkeley.edu/).

SAC Joint Venture. 2000. "Design Criteria for New Moment-Resisting Steel Frame Structures." Report No. SAC-2000-01, Publication Pending.

SEAOC Vision 2000. A Framework for Performance Based Design I, II, & III, Structural Engineers Association of California, SEAOC (1995).

6 ET Analysis of Framed Structures

6.1 INTRODUCTION

In this chapter, application of the Endurance Time (ET) method in the analysis of moment frames will be demonstrated and verified.[1] Moment frames are among the most popular framing systems in buildings and other structures. They are also important from the viewpoint of structural dynamics as a good representative of multi-degrees-of-freedom (MDOF) systems. Quantification of seismic demands for seismic performance assessment of moment frames implies the statistical and probabilistic evaluation of Engineering Demand Parameters (EDPs), that is, story drifts, floor acceleration, and so on as a function of ground motion Intensity Measures (IMs), that is, peak ground acceleration, spectral acceleration at the first-mode period, and so on. Sensitivity of the relationship between EDPs and IMs to important structural and ground motion characteristics should also be studied (Medina and Krawinkler 2005). Several research efforts have focused on the evaluation of demands for both single- and multi-degrees-of-freedom (SDOF and MDOF) systems in which displacement demands from nonlinear response history analyses have been quantified as a function of a normalized strength or ground motion intensity level (Seneviratna and Krawinkler 1997; Whittaker, Constantinou, and Tsopelas 1998; Fajfar 2000; Gupta and Krawinkler 2000; Miranda 1999; Teran-Gilmore 2004; Chopra, Goel, and Chintanapakdee 2003).

Current structural engineering practice estimates seismic demands by the nonlinear static procedure, or "pushover analysis," detailed in Federal Emergency Management Agency (FEMA-356) or Applied Technology Council (ATC-40) guidelines (FEMA-356 2000; ATC-40 1996). The seismic demands are computed by nonlinear static analysis of the structure subjected to monotonically increasing lateral forces with an invariant or variant height-wise distribution until a target value of roof displacement is reached. This roof displacement value is determined from the earthquake-induced deformation of an inelastic single-degree-of-freedom (SDOF) system derived from the "pushover curve" (Chopra, Goel, and Chintanapakdee 2003).

Another promising tool for estimating inelastic deformation demands is Incremental Dynamic Analysis (IDA). In IDA, the seismic loading is scaled and different nonlinear dynamic analyses implemented to estimate the dynamic performance of the global structural system (Vamvatsikos and Cornell 2002). By using this

DOI: 10.1201/9781003217473-6

method, EDPs of the structures can be obtained at different IMs and therefore the performance of the structures can be reviewed more precisely. A large number of nonlinear dynamic analyses are needed for the IDA method and various performances of the structure to the different records applied to it are the main drawbacks of this method in practical application.

The Endurance Time (ET) method is basically a simple dynamic pushover test that tries to predict EDPs of structures at different IMs by subjecting them to some predesigned intensifying dynamic excitations. These predesigned excitations in the ET method are called "acceleration functions" or, more generally, "excitation functions" in order to clearly identify them from ground motions and simulated accelerograms that are usually compatible with ground motions. The ET acceleration functions are designed in a manner that their intensity increases with time. In order to practically apply the ET method as a tool for design and assessment of structures, ET acceleration functions should preferably represent different earthquake hazard levels at different times as far as possible. For this purpose, the concept of response spectra has been taken advantage of in developing ET acceleration functions as discussed in previous chapters (Estekanchi, Valamanesh, and Vafai 2007; Estekanchi, Arjomandi, and Vafai 2007). As previously explained, numerical optimization techniques are used in order to create a set of ET acceleration functions with the property of having a response spectra that proportionally intensifies with time while remaining compatible to a prespecified target response spectra curve. Detailed procedure for generating ET acceleration functions was described in Chapter 4. Because of the increasing demand for the ET acceleration function, structures gradually go through elastic to yielding and nonlinear inelastic phases, finally leading to global dynamic instability.

Some early studies have shown that in linear seismic analysis of structures, the ET method can reproduce the results of codified static and response spectrum analysis procedures with acceptable accuracy (Estekanchi, Arjomandi, and Vafai 2007). Compliance and level of accuracy of this method in nonlinear seismic analysis of SDOF structures has also been investigated. In this chapter, accuracy and consistency of the ET method in estimating average inelastic deformation demands of regular steel frames will be the focus of discussion. These studies are required in order to provide a basis for practical application of the ET method in steel frames seismic assessment and design problems. To reach this goal a set of ground motions is selected, and their average response spectrum is calculated. This spectrum is set to be the target response spectrum for generating a set of acceleration functions used in this study (i.e., ETA20f series). A set of steel moment-resisting frames with different number of stories was used in this study. This set consists of under-designed, adequately designed, and over-designed frames to examine the capability of the ET method in differentiating dissimilar structures. Elastic-Perfectly-Plastic (EPP) material model and bilinear material model with a postyield stiffness equal to 3 percent of the initial elastic stiffness (STL) are used to study the nonlinear behavior of the frames. The results computed with the ET method were compared to the results of nonlinear response history analyses. A procedure is described to find an equivalent time in the ET analysis to compare its results with the results of nonlinear response history analysis. Mean values and dispersions of the results obtained by two methods are compared for different frames.

Finally, potential application of the ET method for seismic rehabilitation of structures is explained by using dampers at different stories of a sample structure.

For frames with the EPP material model, which are P-Δ sensitive cases, estimations of ET analysis for maximum interstory drift ratio are less than non-linear history analysis results. Nonlinear response history analysis of flexible structures that are subjected to large displacements may be severely influenced by the P-Δ effects. For these cases, the maximum interstory drift ratio becomes very sensitive to ground motions that are relatively strong. Therefore, for these ground motions, P-Δ effects destabilize the structure and increase the maximum interstory drift ratio, drastically resulting in the average value of this parameter to become unreliable and scattered. As will be shown later, average values of deformation demands cannot be reliably predicted by applied ET acceleration functions in these cases. Unlike real earthquakes, ET acceleration functions used in this study have quite similar characteristics. Consequently, the dispersion of the results of nonlinear response history analysis for these frames is high. However, it should be noted that the deformations resulting from the destabilizing effect of P-Δ in EPP models are usually beyond the drift levels that are practically important for design and are of little significance. The consistency between the results of ET analysis and nonlinear response history analysis for the material model with 3 percent postyield stiffness is satisfactory. These frames are much less sensitive to P-Δ effects when subjected to strong ground motions. The consistency of the base shears obtained by two methods is acceptable for both material models. As will be shown, ET can be considered as a useful approximate analysis procedure that provides a practical tool by drastically reducing the number of required non-linear time-history analyses at each step of design refinement.

6.2 PROCEDURE FOR APPLICATION OF THE ET METHOD

In the ET method, numerical (or experimental) models of structures are subjected to intensifying acceleration functions. Major structural responses, such as displacements, drift ratios, stresses, or other appropriate EDPs are monitored up to the desired limiting point where the structure collapses or failure criteria are met. Each specific time in ET analysis can be correlated to a specific Intensity Measure (IM) that expresses an earthquake hazard level. The details of application and interpretation of the results from the ET method can be somewhat different. In ET analysis for seismic performance assessment, which will be explained here, the equivalent time for each seismic hazard level can be defined, and the performance of the structure until that time can be compared by a predefined performance objective. Based on each different structural performance level, such as Immediate Occupancy (IO), Life Safety (LS) and Collapse Prevention (CP), and the design or rehabilitation objectives, corresponding equivalent time can be defined and alternative designs compared at these milestone times representing different excitation levels in each single analysis. Basically, the longer the structure can endure imposed excitations, it is judged to have better performance. In practice, the analysis or experiment need not be continued until the real collapse of the structure. Any convenient performance parameter, such as maximum drift, stress ratio, plastic rotation, can be considered, and analysis or experiment can be commenced until the desired level of excitation has been covered (Estekanchi, Vafai, and Sadeghazar 2004).

One of the most important issues in successful implementation of the procedure is determination of a suitable acceleration function so that the results from ET analysis (or testing) can be correlated reliably well with the response of structures subjected to earthquakes. The concept of response spectra can be used in producing the intensifying ET acceleration functions for this purpose (Estekanchi, Valamanesh, and Vafai 2007; Estekanchi, Arjomandi, and Vafai 2007). However, production of ET acceleration functions is an area of active research, and there are various approaches that can be adopted for their production (Estekanchi et. al. 2020). Optimization techniques are normally used in order to create a set of ET acceleration functions with response spectrum that proportionally intensifies with time while remaining compatible to a prespecified template response spectra as far as possible. This means that the response spectrum of any window of these acceleration functions from $t_0 = 0$ to $t_1 = t$ resembles that of the target spectrum with a scale factor that is proportional with time (t) (Estekanchi, Valamanesh, and Vafai 2007). Even though other strategies can also be used to define intensifying ET acceleration functions, the stated method seems to provide a well-suited acceleration function for the purposes of this study. To apply the ET method for Performance Based Seismic Engineering (PBSE), it is suitable to generate acceleration functions that their response spectra are compatible with the response spectra of different hazard levels. Generally, the seismic hazard due to ground shaking is defined for any earthquake hazard level using spectral response acceleration. The response spectra for different hazard levels can be used for the generation of ET acceleration functions. The optimization procedure for generating ET acceleration functions is drastically time consuming but once a set of acceleration functions based on earthquake hazard levels of a seismic code is generated, it can be used easily for any structure.

Although current sets of ET acceleration functions are generated based on linear response spectra, their performance in estimating nonlinear response of SDOF systems has been satisfactory. In this chapter, the accuracy of ET analysis in estimating average inelastic deformation demands is examined by comparing its results with nonlinear response history analysis results. The response spectra used for the generation of ET acceleration functions are representative of the average response of the structures subjected to a set of earthquakes. ET analysis in this study is aimed to predict this average response. For design purposes, the response spectrum function can be adjusted to properly reflect the level of dispersion by applying statistical procedures. ET acceleration functions can then be generated based on these design spectrums. In this study, however, the average response spectrum from seven earthquakes is directly used in order to investigate the accuracy and consistency of ET analysis and directly compare them with the results of traditional time-history analysis.

ET acceleration functions that have been used in this research are designed in such a way that their response spectrum remains proportional to that of the average of seven strong motions recorded on a stiff soil condition. These acceleration functions are generated to be compatible with some ground motions to facilitate the comparison of the results of ET analysis and nonlinear response history analysis. The dynamic properties of these acceleration functions will be discussed in the next sections.

6.3 SPECIFICATIONS OF MODELS, GROUND MOTIONS, AND ACCELERATION FUNCTIONS

A set of steel moment-resisting frames with different numbers of stories are studied. This set consists of two-dimensional regular generic frames with 3, 7, and 12 stories. These frames have either a single bay or three bays. Generic frames of this study are based on the models developed by Estekanchi et al. (Estekanchi, Valamanesh, and Vafai 2007). These frames are designed according to AISC-ASD design code (American Institute of Steel 1989). To facilitate the comparative studies, frames are designed for different levels of lateral loads, named as "Standard," "Under-designed," and "Over-designed." Standard frames have been designed according to the recommendations of the INBC for a high seismicity area (BHRC 2005). Frames are designed with the aid of equivalent static procedure. The name of these frames ends with the letter S (Standard). Under-designed frames have been designed assuming one-half of the codified base shear as the design lateral load. The name of these frames ends with the letter W (Weak). Over-designed frames have been designed for twice the standard lateral load. The name of these frames ends with the letter O (Over-designed). Frame masses are considered to be the same for these three kinds of frames. A response modification factor of $R = 6$ is used for the design of the frames, and interstory drift ratio is limited to 0.005 for all of them per code requirements. Usually, in standard and over-designed frames, story drift turns out to be the controlling design criteria. In under-designed frames, elemental forces control the design. Geometry and section properties of the frames with seven stories and one bay are depicted in Figure 6.1. Some of the basic specifications of all frames are shown in Table 6.1.

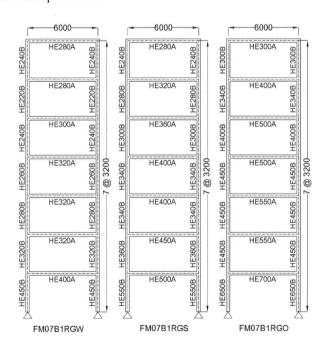

FIGURE 6.1 Schematics of the frames with seven stories and one bay.

TABLE 6.1
Specifications of the Frames

Frames	Number of stories	Number of bays	Mass participation Mode 1	Fundamental period (sec)	Design base shear (KN)
FM03B1RGW	3	1	91.0%	1.20	59.7
FM03B1RGS	3	1	88.0%	0.89	116
FM03B1RGO	3	1	85.1%	0.60	245
FM03B3RGW	3	3	88.6%	1.25	179
FM03B3RGS	3	3	85.7%	0.89	362
FM03B3RGO	3	3	85.6%	0.61	729
FM07B1RGW	7	1	81.2%	2.03	101
FM07B1RGS	7	1	80.6%	1.43	205
FM07B1RGO	7	1	80.6%	0.99	415
FM07B3RGW	7	3	81.3%	2.05	302
FM07B3RGS	7	3	80.9%	1.44	610
FM07B3RGO	7	3	80.4%	0.97	1,233
FM12B3RGW	12	3	79.3%	2.89	399
FM12B3RGS	12	3	78.4%	2.05	804
FM12B3RGO	12	3	75.2%	1.30	1,631

Elastic-Perfectly-Plastic material model and bilinear material model with a post-yield stiffness equal to 3 percent of the initial elastic stiffness are used to study the nonlinear behavior of the frames. An EPP model has been used widely in previous investigations and therefore it represents a benchmark to study the effect of hysteretic behavior. Furthermore, recent studies have shown that this is a reasonable hysteretic model for steel beams that do not experience lateral or local buckling or connection failure (Foutch and Shi 1998). For more realistic non-linear behavior, an STL material model is also used in this study. To apply these material models in the analysis the OpenSees beam-column element with non-linear distributed plasticity is utilized (OpenSees 2002). This element is used for beams and columns of the frames to account for the nonlinearity for both of them. Only one (horizontal) component of the ground motion has been considered while dynamic soil-structure interaction is neglected. P–Δ effects have been included in the analysis. A viscous damping of 5 percent, as customary for these types of frames, has been applied in the analyses.

To investigate the accuracy of the ET method in estimating nonlinear response of ground motions, a set of ET acceleration functions (ETA20f) are used that are consistent with the average response spectrum of ground motions. To reach this goal, 20 accelerograms that are recorded on Site Class C, as defined by the NEHRP, and used in FEMA 440 were selected (FEMA-440 2005). From these ground motions, seven records that their response spectra shape were more compatible with the response spectrum of soil type II of INBC standard 2800 were selected (Table 6.2) (Riahi and

TABLE 6.2
Description of GM1 Set of Ground Motions Used in This Study

Date	Earthquake name	Magnitude (Ms)	Station number	Component (deg)	PGA (cm/s²)	Abbreviation
06/28/92	Landers	7.5	12149	0	167.8	LADSP000
10/17/89	Loma Prieta	7.1	58065	0	494.5	LPSTG000
10/17/89	Loma Prieta	7.1	47006	67	349.1	LPGIL067
10/17/89	Loma Prieta	7.1	58135	360	433.1	LPLOB000
10/17/89	Loma Prieta	7.1	1652	270	239.4	LPAND270
04/24/84	Morgan Hill	6.1	57383	90	280.4	MHG06090
01/17/94	Northridge	6.8	24278	360	504.2	NRORR360

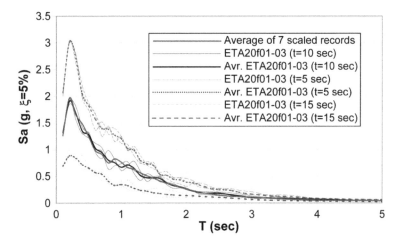

FIGURE 6.2 Total acceleration response spectra of ETA20f series acceleration functions for x=5% at different time.

Estekanchi 2006). These seven accelerograms are scaled to produce response spectrum compatible with INBC standard 2800 spectrum. Finally, an average of pseudo acceleration spectrum of these scaled accelerograms was obtained and smoothed. The smoothed spectrum is used as the target spectrum in generating the ETA20f series of ET acceleration functions used here. As can be seen in Figure 6.2, response spectrum of a window of ET acceleration functions from $t_0 = 0$ to $t_1 = 10$, that is, $t \in [0, 10]$, matches reasonable well with the average response spectrum of the seven strong motion records.

It is important to note that the ET response spectra remains proportional to the target spectra from seven ground motions at all times – for example, it is 0.5 and 1.5 times the target spectra at $t = 5$ sec and $t = 15$ sec respectively. A sample acceleration function generated in this way is shown in Figure 6.3. To compare the results of ET

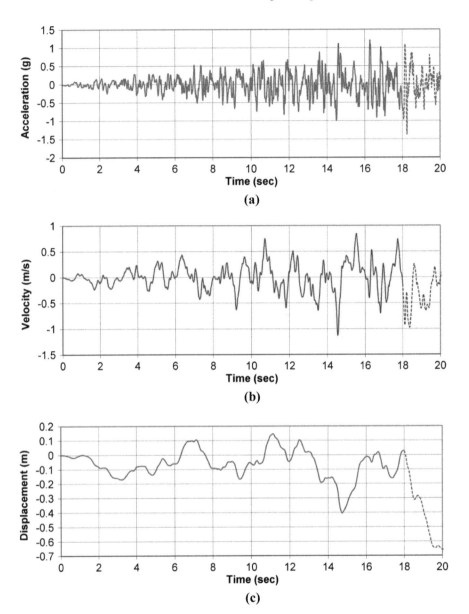

FIGURE 6.3 (a) ETA20f02 acceleration function; (b) ETA20f02 velocity function; (c) ETA20f02 displacement function.

analysis with earthquakes, a set of ground motions (GM1) are used. The set consists of seven records that are also used for the generation of ETA20f set of acceleration functions.

To be consistent with the seismic codes, a GM1 set of ground motions should be scaled. All frames are analyzed as planar structures subjected to a single horizontal component of ground motion. Therefore, records are scaled individually rather than as pairs. Ground motions are scaled such that their average 5 percent-damped linear spectra does not fall below the design spectrum in the periodic range of $0.2T_i$ to $1.5T_i$, where T_i is the fundamental period of vibration of the frame. Here scale factors are obtained in a way that the ground motion spectrum matches the ASCE-7 spectrum in the mentioned range (ASCE 2006). Scale factors obtained by this method for a GM1 set are shown in Table 6.3 for each frame.

6.4 COMPARISON WITH NONLINEAR RESPONSE HISTORY ANALYSIS

For each set of ground motions and acceleration functions, mean value and standard deviation of the specified damage measure can be calculated. For a set of ground motions, for example, mean value and standard deviation of the EDP can be calculated by the following equation:

$$\overline{EDP}_{ex} = \frac{1}{n}\sum_{i=1}^{n}EDP_{ex,i} \tag{6.1}$$

$$\sigma_{ex} = \sqrt{\frac{\sum_{i=1}^{n}(EDP_{ex,i} - \overline{EDP}_{ex})^2}{n-1}} \tag{6.2}$$

In this equation $EDP_{ex,i}$ is the value of EDP for a ground motion, n is the number of ground motions in the set, $\overline{EDP}_{ex}$ is the mean value of EDP for the set, and σ_{ex} is the standard deviation of it. Similar values can be calculated for the set of acceleration functions.

An important question is how the results of two methods can be compared. Results of ET analysis are obtained through time and, as mentioned before, in this method time is correlated with IM. Therefore, different values of EDP are calculated for different values of IM in an ET analysis. To establish a relation between the results of the ET method and any other method, the IM value of the other method should be found in ET analysis. Therefore, a procedure should be adopted to find an equivalent time in ET analysis in which the IM values of the two methods are equal.

Many quantities have been proposed to characterize the intensity of a ground-motion record. In the ET method, intensity increases through time and, therefore, scalable IMs can be conveniently used. Common examples of scalable IMs are the Peak Ground Acceleration (PGA), Peak Ground Velocity (PGV), and the spectral acceleration at the structure's first-mode period $(S_a(T_j))$. Here, first-mode spectral acceleration $(S_a(T_j))$ is used as an IM to obtain the equivalent time. Most of the frames used in this study are first-mode dominated structures that are sensitive to the strength of the frequency content near their first-mode frequency, which is well characterized by $S_a(T_j)$. Moreover, $S_a(T_j)$ produces a relatively low dispersion over the full range of

TABLE 6.3
Scale Factors of GM1 Set for Different Frames

Frames	Scale factors								Equivalent time (sec)
	LPAND270	LADSP000	MHG06090	LPGIL067	LPLOB000	NRORR360	LPSTG000		
FM03B1RGW	2.89	3.97	1.74	2.35	2.63	1.11	1.61		10.79
FM03B1RGS	2.55	3.66	1.61	2.12	1.87	1.15	1.76		10.26
FM03B1RGO	2.36	3.64	1.79	1.91	1.60	1.31	1.87		9.63
FM03B3RGW	2.92	4.02	1.77	2.40	2.78	1.11	1.60		10.96
FM03B3RGS	2.61	3.68	1.61	2.14	1.93	1.14	1.75		10.36
FM03B3RGO	2.36	3.62	1.72	1.99	1.68	1.20	1.87		9.48
FM07B1RGW	3.51	4.58	2.30	3.15	4.59	1.25	1.59		13.13
FM07B1RGS	3.08	4.18	1.92	2.58	3.29	1.13	1.55		11.47
FM07B1RGO	2.76	3.75	1.62	2.20	2.14	1.10	1.71		10.64
FM07B3RGW	3.63	4.70	2.35	3.41	5.05	1.27	1.59		13.51
FM07B3RGS	3.13	4.22	1.96	2.65	3.49	1.13	1.54		11.67
FM07B3RGO	2.69	3.71	1.61	2.17	2.04	1.12	1.72		10.57
FM12B3RGW	4.26	5.46	2.82	4.06	6.24	1.49	1.78		15.36
FM12B3RGS	3.57	4.64	2.33	3.28	4.81	1.26	1.59		13.28
FM12B3RGO	2.96	4.07	1.82	2.47	2.99	1.12	1.57		11.16

EDP values in IDA analysis (Vamvatsikos and Cornell 2002). Certainly, other IMs can easily be used to calculate the equivalent time. The equivalent time can be calculated for a single record or a set of records. To compare the results of a set of records with the results of ET analysis, the average of the first-mode spectral acceleration of the records ($S_{a,Ave}$) can also be calculated. Furthermore, the value of the smooth response spectrum used for the generation of ET acceleration functions at the first-mode period (T_1) is calculated ($S_{a,ET}$). Finally, the equivalent time is obtained by Equation 6.3.

$$t_{eq} = \frac{S_{a,Ave}}{S_{a,ET}} \times 10 \qquad (6.3)$$

Constant 10 is used in this equation because the response spectrum of ET acceleration function at $t = 10$ sec matches the target smooth response spectrum with a scale factor of unity. Equivalent times of each frame for GM1 set of records obtained by this procedure are shown in Table 6.3.

As can be seen in Table 6.3, for frames that have long periods, the equivalent time is greater than 10 seconds, which is the target time in generation of ETA20f series of acceleration functions. The main reason of obtaining such a large equivalent time is the scaling procedure used for the records. For example, the equivalent time for FM12B3RGW frame is 15.36 seconds. The period of this frame is 2.89 seconds and therefore the scaling procedure is done for the range of 0.578 to 4.335 seconds. In this range, the smooth spectrum used for the generation of ET acceleration functions is lower than the ASCE-7 spectrum (Figure 6.4). Therefore, large-scale factors are obtained for this frame (Table 6.3). It can be concluded that for a better comparison of the results of ET analysis with the results of other methods, the response spectrum used for the generation of acceleration functions should be consistent with the spectrum used in other methods in a wide range of periods (e.g., 0 to 5 seconds).

All of the frames have been subjected to the ETA20f set of acceleration functions and GM1 set of ground motions. The response data summarized in this research are part of a comprehensive database on EDPs acquired for the previously defined generic frames. The discussion presented here focuses on maximum interstory drift ratios for frames. This EDP is relevant to structural damage if damage is dominated by the maximum story deformation over the height and is a measure of damage to nonstructural components. For P-Δ sensitive structures, maximum interstory drift ratio is the most relevant EDP for global collapse assessment because dynamic instability is controlled by the story in which the story drift grows most rapidly (Medina and Krawinkler 2005). To examine the consistency of the IM of nonlinear response history analysis and IM obtained from ET analysis, base shear of the frames obtained by two methods are compared.

The average of maximum interstory drift ratios for ET acceleration functions through time and values of equivalent time are presented for FM03B1RG frames in Figure 6.5. These results are obtained considering the EPP material model for the frames. It should be noted that ET analysis results are usually presented by increasing

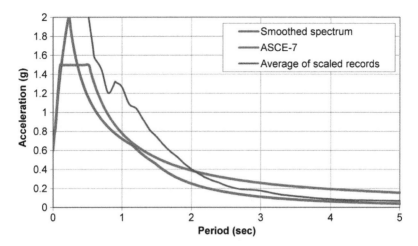

FIGURE 6.4 Comparison of smooth spectrum and ASCE-7 spectrum with the average spectrum of the scaled records for FM12B3RGW frame.

curves, where the y coordinate at each time value t, corresponds to the maximum absolute value of the required parameter in the time interval $[0, t]$ as given in Equation 6.4.

$$\Omega\big(f(t)\big) \equiv Max(Abs\big(f(\tau): \tau \in [0,t]\big)) \qquad (6.4)$$

Here Ω is the Max-Abs operator, as defined above, and $f(t)$ is the response history such as base shear, interstory drift, damage index, or other parameters of interest.

Because of the statistical characteristics and dispersion of ET analysis results in nonlinear range, the resulting curves are serrated. Sometimes the value of the response does not pass the maximum value experienced previously in a long-time interval. Therefore, the resulting ET curve has a constant value in that interval. In this research, a moving average procedure is used to reduce the serrated nature of the ET curves in nonlinear range.

A typical ET analysis curve can reveal significant information about a structure. During the initial phase of the excitation, the structure behaves linearly until it reaches a certain point that a plastic hinge is created – that is, plastic behavior. By increasing the intensity of the acceleration function through time, the structure experiences more plastic deformations until it reaches the collapse limit. Occurrence of the collapse for a structure through an ET analysis is mostly dependent on its lateral stiffness and strength. For example, FM03B1RGW frame experiences significantly high displacements after 12 seconds. But the two other frames do not collapse before the 20th second, as can be seen in the figure.

Maximum interstory drift ratios of FM03B1RG frames with the EPP material model obtained from nonlinear response history analysis are presented in Figure 6.6. As it can be seen in Figure 6.6, all of the records rank the structures based on their lateral stiffness. This result can be seen in ET curves, too. Maximum interstory drift ratios of

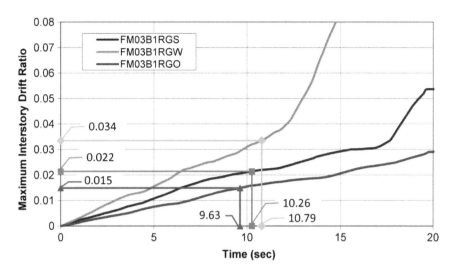

FIGURE 6.5 ET maximum interstory drift ratio curves for FM03B1RG frames with EPP material model and values of t_{eq} of nonlinear response history analysis.

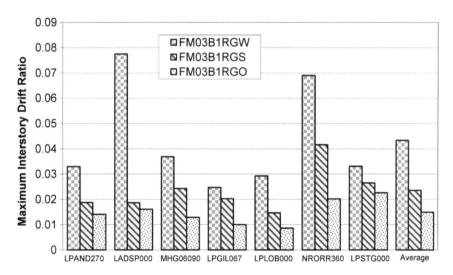

FIGURE 6.6 Maximum interstory drift ratios of the accelerograms and their average for FM03B1RG frames with EPP material model.

FM03B1RGW frame for NRORR360 and LADSP000 records are far beyond the average values of other records. In nonlinear response history analysis, such exceptions can significantly affect the average value of the response. It will be discussed later that the main cause of the differences between the results is the P-Δ effect. As can be seen in Table 6.4, the ET method underestimates the results of this frame. This means that ET analysis with ETA20f series could not estimate these exceptions well.

TABLE 6.4

Comparison between the Results of Nonlinear Response History Analysis and ET Analysis for Different Frames with EPP Material Model

Frames	Maximum interstory drift ratio				Base shear (KN)	
	Average		Standard deviation		Average	
	THA	ETA	THA	ETA	THA	ETA
FM03B1RGW	0.043	0.034	0.021	0.004	218.1	217.8
FM03B1RGS	0.024	0.021	0.009	0.003	323.0	335.6
FM03B1RGO	0.015	0.015	0.005	0.001	505.3	558.3
FM03B3RGW	N.A.	0.057	N.A.	0.022	N.A.	543.6
FM03B3RGS	0.045	0.025	0.037	0.003	912.5	933.7
FM03B3RGO	0.017	0.014	0.007	0.002	1,424.7	1,513.3
FM07B1RGW	0.032	0.026	0.011	0.003	284.9	286.6
FM07B1RGS	0.023	0.022	0.004	0.005	433.6	446.5
FM07B1RGO	0.015	0.014	0.002	0.002	729.3	733.2
FM07B3RGW	N.A.	0.037	N.A.	0.007	N.A.	738.1
FM07B3RGS	0.025	0.023	0.003	0.005	1,256.4	1,274.4
FM07B3RGO	0.016	0.015	0.003	0.003	2,098.2	2,109.0
FM12B3RGW	0.032	0.025	0.004	0.004	1,002.5	991.6
FM12B3RGS	0.029	0.022	0.010	0.002	1,465.1	1,481.6
FM12B3RGO	0.015	0.015	0.003	0.001	2,898.1	2,983.1

Figures 6.7 and 6.8 show the interstory drift ratio response history of the FM03B3RGW frame for LPAND270 record and ETA20f02 acceleration function, respectively. Most of the records, and all of the ET acceleration functions, anticipate that the maximum interstory drift ratio is observed in second story for this frame. The difference between the results of this story and the others is significant in high IMs.

As another example, the average of maximum interstory drift ratios for ET acceleration functions through time are presented for FM07B1RG frames with EPP material model in Figure 6.9. As can be seen, the general trend of the results are similar to FM03B1RG frames results, with some exceptions. ET results show that between $t = 9$ and 13 sec the maximum interstory drift ratio of the FM07B1RGW frame is less than what is obtained for the FM07B1RGS frame. This estimation of the ET method can be checked by nonlinear response history analysis. To do so, a reduction-scale factor is applied to GM1 set to change the equivalent time of the FM07B1RGW frame to the equivalent time of the FM07B1RGS frame. Finally, the FM07B1RGW frame is analyzed for the GM1 set with this reduction-scale factor. Now the results of the FM07B1RGW and FM07B1RGS frames can be compared at $t_{eq} = 11.47$ sec. At this time, the maximum interstory drift ratio of the FM07B1RGS frame obtained from ET analysis is larger than the corresponding value of the FM07B1RGW frame. But the results of nonlinear response history analysis are vice versa. The maximum interstory drift ratio of the FM07B1RGS frame is 0.0226, and the corresponding value for the

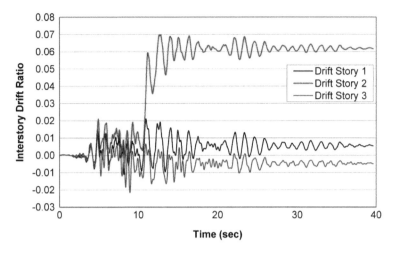

FIGURE 6.7   Interstory drift ratio response history of FM03B3RGW frame for LPAND270.

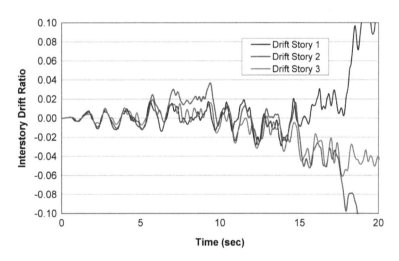

FIGURE 6.8   Interstory drift ratio response history of FM03B3RGW frame for ETA20f02.

FM07B1RGW frame at t_{eq} = 11.47 sec is 0.024. It shows that when the EDPs of the frames are very close together, small differences in ET analysis curves might be a randomness effect and should not be interpreted as an indication that one structure has better performance over the other. In these cases, performance differences may actually be insignificant. A more refined analysis using a higher number of ground motions and improved ET acceleration functions are required if a definitive conclusion is to be made in such cases.

Figure 6.10 shows the average of maximum interstory drift ratios of FM07B3RG frames with EPP material model for the ET acceleration functions through time. Like the previous examples, ET curves differentiate between three frames. Between t = 18 and 20 sec, the maximum interstory drift ratios of the FM07B3RGO frame are

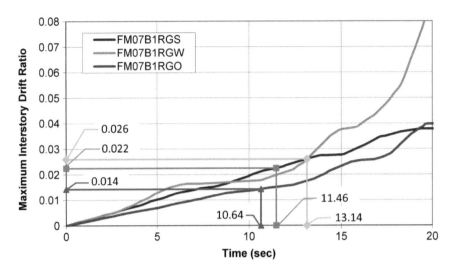

FIGURE 6.9 ET maximum interstory drift ratio curves for FM07B1RG frames with EPP material model and values of t_{eq} of nonlinear response history analysis.

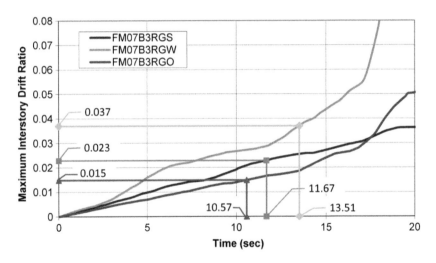

FIGURE 6.10 ET maximum interstory drift ratio curves for FM07B3RG frames with EPP material model and values of t_{eq} of nonlinear response history analysis.

larger than the corresponding value for the FM07B3RGS frame. This can be verified by nonlinear response history analysis for the equivalent IM for this range of time. To do so, increasing scale factors are applied to the GM1 set for the FM07B3RGO and FM07B3RGS frames. These scale factors change the equivalent time of these frames to 20 seconds. The maximum interstory drift ratios of FM07B3RGS and FM07B3RGO frames at t_{eq} = 20 sec are 0.0444 and 0.0663, respectively. This time the results of nonlinear response history analysis are consistent with the results of ET analysis.

Average and standard deviation of maximum interstory drift ratios and average of base shears obtained from nonlinear response history analysis and ET analysis of the frames with EPP material model are compared in Table 6.4. It should be noted that for two cases of under-designed frames, nonlinear response history analysis of some ground motions did not converge and, therefore, no results are presented for them. In most of the frames, estimations of ET analysis for maximum interstory drift ratio are less than nonlinear response history analysis results. The difference between the results is more in under-designed frames, which experience more nonlinearity in their analyses. Usually, in these frames, the dispersion of the results of nonlinear response history analysis is high. But in frames that behave more linearly than others, the difference between the results of ground motions is less, and the results match the results of ET analysis better.

Table 6.4 shows that the consistency of the base shears obtained by two methods is acceptable. It means that the procedure to find the equivalent time in ET analysis to match the IMs of two methods works well. It should be noted that although the equivalent time tries to make a better consistency between the IMs of two methods, it is done just for one period. Therefore, the average response spectrum of ET acceleration functions at $t = t_{eq}$ and the average response spectrum of scaled accelerograms have some minor differences that cause the inconsistency of the base shears in some frames like FM03B1RGO and FM03B3RGO. The difference between the base shears obtained by two methods is more significant in over-designed frames and in over-designed and properly designed frames, the ET method always overestimates the base shear.

Figure 6.11 shows the mean and mean plus and minus one standard deviation (STDEV) of maximum interstory drift ratios of the frames with the EPP material model obtained by nonlinear response history analysis. The average of maximum

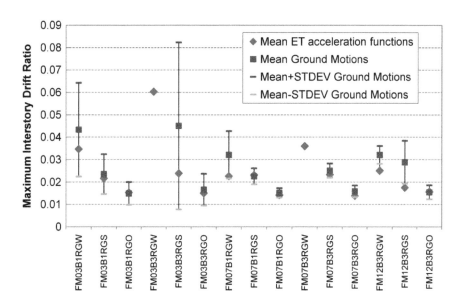

FIGURE 6.11 Comparison of average maximum interstory drift ratios of the frames with EPP material model obtained by nonlinear response history analysis and ET analysis.

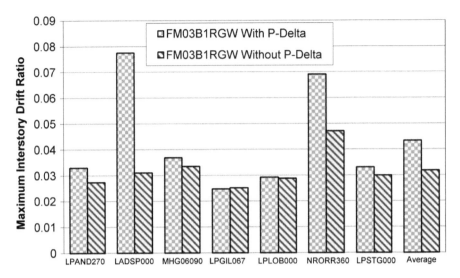

FIGURE 6.12 Maximum interstory drift ratios of the accelerograms and their average for FM03B1RGW frame with EPP material model with and without considering P-Δ effects.

interstory drift ratios of the frames obtained by ET analysis are also shown in this figure. If this figure is compared with Table 6.4, it can be concluded that, when the dispersion of the results is high, and the difference between the results of the two methods is also high. For example, Figure 6.11 shows that the dispersion of the results of under-designed frames is larger than properly designed and over-designed frames. For these frames, Table 6.4 and Figure 6.11 show the maximum difference between interstory drift ratios obtained by two methods. For the FM03B3RGS frame, the dispersion is high, and the difference between the results of the two methods is also high.

The P-Δ effect increases the dispersion of the results of nonlinear response history analysis. Nonlinear seismic response of steel moment-resisting frame structures, which are usually quite flexible, may be severely influenced by the structure P-Δ effect. This especially occurs when these structures are subjected to large displacements under severe ground motions. For structures in which this effect induces negative postyield story stiffness, the responses become very scattered under severe ground motions (Gupta and Krawinkler 2000).

Figure 6.12 compares the results of nonlinear response history analysis of the FM03B1RGW frame with the EPP material model by considering and eliminating the P-Δ effect. In most of the accelerograms, the maximum interstory drift ratio obtained for the model eliminating the P-Δ effect is less than the model considering this effect. The largest difference between these results is obtained for LADSP000 and NRORR360 records. Also, the responses of the frame to these records are the largest in the GM1 set. By eliminating the P-Δ effect, the results of these records approaches the average response for the GM1 set, and the dispersion of the results reduces significantly.

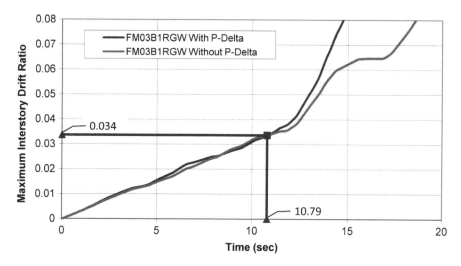

FIGURE 6.13 ET maximum interstory drift ratio curves for FM03B1RGW frame with EPP material model with and without considering P-Δ effects.

The same trend can be seen in the results of ET analysis. Figure 6.13 compares the results of ET analysis of the FM03B1RGW frame with the EPP material model by considering and eliminating the P-Δ effect. It can be seen that the curves are separated from each other at about $t = 11$ sec. After this time the P-Δ effect increases the maximum interstory drift ratio of the frame. As it can be seen in the figure, the P-Δ effect changes the EDP after the equivalent time computed for nonlinear response history analysis. In other words, the effects of P-Δ on the results of ET analysis and nonlinear response history analysis are not seen at the same IM. The reason for this phenomenon is the difference in the nature of acceleration functions and ground motions. For some of the ground motions, P-Δ effects drastically increase the maximum interstory drift ratio, and the average value of this parameter is changed considerably. This is due to the fact that the characteristics of the ground motions can be very different from each other. Unlike actual earthquakes, ET acceleration functions have similar characteristics. Consequently, the dispersion of the results of nonlinear response history analysis for these frames is high but, in ET analysis, the dispersion of the results is not significant.

P-Δ effects are not critical if the effective stiffness at maximum displacement remains positive (Gupta and Krawinkler 2000). As mentioned, the EPP material model was used for previous models. Because this material model does not have strain hardening, the frames tend to reach negative postyield stiffness, especially in under-designed frames and, therefore, P-Δ effects can change their drifts considerably. If the STL material model, which has a 3 percent postyield stiffness, is used instead of the EPP material model, it can be guessed that the results of the ET analysis approach the nonlinear response history analysis results.

Figure 6.14 compares the ET maximum interstory drift ratio curves for FM03B1RG frames with EPP and STL material models. As can be seen by increasing

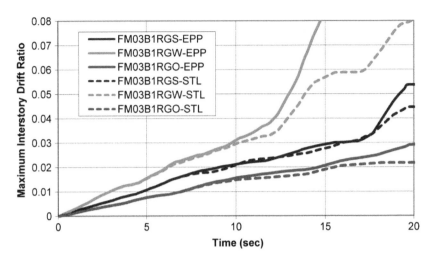

FIGURE 6.14 ET maximum interstory drift ratio curves for FM03B1RG frames with EPP and STL material model.

the strain-hardening rate of the material model, the P-Δ effects are decreased. If this figure is compared with Figure 6.13, it can be concluded that ET results of the FM03B1RGW frame with the STL material model with respect to P-Δ effects are very similar to the results of this frame with the EPP material model eliminating P-Δ effects. Average and standard deviation of maximum interstory drift ratios and the average of base shears obtained from nonlinear history analysis and ET analysis of the frames with STL material model are compared in Table 6.5. If this table is compared with Table 6.4, it can be seen that with regard to 3 percent strain hardening, the differences between the results of the two methods are totally decreased. It can be concluded that, for structures that are extremely sensitive to P-Δ effects, the results of ET analysis should be used with special care. To avoid underestimated values for EDPs – such as maximum interstory drift ratio – it is better to define a limit for the lateral deformation of the structure. This limit should specify the onset of reaching negative postyield story stiffness. Interstory drift ratio obtained at the maximum base shear of a pushover curve can be a good reference value for this limit.

The maximum interstory drift ratios and corresponding base shears are calculated at t_{eq}, but these values do not necessarily happen at the same analysis time. ET curves are obtained by the Max-Abs operator described in Equation 6.4, and perhaps the maximum value for the base shear is not obtained at the same time that the maximum value of maximum interstory drift ratio is obtained.

The results of ET analysis reported here can be used in order to estimate the mean demand structures due to ground motions. However, the scatter of the results from earthquakes is not anticipated by the ET method. Therefore, a safety margin for expecting seismic demands should be defined. In spectral analysis, this issue is taken into account by using mean spectrum plus standard deviation of the ground motions. The same method can also be used in ET analysis. In addition to generating ET acceleration functions based on a mean spectrum, another set of acceleration functions can be generated by assuming mean spectrum plus standard deviation as the target spectrum. The results of ET analysis

TABLE 6.5
Comparison between the Results of Nonlinear Response History Analysis and ET Analysis for Different Frames with STL Material Model

	Maximum interstory drift ratio				Base shear (KN)	
	Average		Standard deviation		Average	
Frames	THA	ETA	THA	ETA	THA	ETA
FM03B1RGW	0.034	0.031	0.006	0.003	227.6	230.8
FM03B1RGS	0.022	0.022	0.006	0.004	339.4	355.4
FM03B1RGO	0.014	0.014	0.004	0.001	523.9	568.3
FM03B3RGW	0.035	0.030	0.009	0.002	590.463	588.9
FM03B3RGS	0.025	0.022	0.006	0.003	964.0	1,019.9
FM03B3RGO	0.015	0.014	0.005	0.002	1,538.9	1,579.8
FM07B1RGW	0.025	0.023	0.003	0.002	294.8	305.6
FM07B1RGS	0.021	0.020	0.003	0.004	470.7	466.0
FM07B1RGO	0.015	0.014	0.001	0.001	779.1	779.6
FM07B3RGW	0.036	0.033	0.006	0.003	790.679	795.9
FM07B3RGS	0.021	0.019	0.003	0.003	1,338.8	1,322.6
FM07B3RGO	0.015	0.014	0.002	0.001	2261.4	2,295.8
FM12B3RGW	0.026	0.022	0.002	0.001	1,036.0	1,067.4
FM12B3RGS	0.022	0.019	0.004	0.001	1,545.8	1,619.9
FM12B3RGO	0.015	0.014	0.002	0.002	2,917.5	3,225.1

obtained for this set can be used as an upper estimation of the results of ground motions, and it can be addressed as a safety margin in ET analysis.

6.5 APPLICATION OF THE ET METHOD IN SEISMIC REHABILITATION OF BUILDINGS

The analysis of under-designed, properly designed, and over-designed frames discussed previously was for explanatory purposes, and their relative performances could be guessed without advanced analysis. In order to further demonstrate the significance of ET analysis, the capability of the ET method in a more complicated situation is demonstrated in this section.

As shown in Table 6.4, the maximum interstory drift ratios of the FM03B3RGW frame with the EPP material model obtained by ET analysis is 0.057. Nonlinear response history analysis of this frame does not converge for two ground motions, and it can be judged that the performance of this frame is not acceptable. Now let us assume that the maximum acceptable interstory drift ratio for this frame has been set to 0.04. In order to improve the frame's performance, a viscoelastic damper is to be used. The question to be answered is in which story level a damper with the damping constant of 2,000 kN.s/m should be installed to result in the best performance – that is, the least overall maximum interstory story drift ratio.

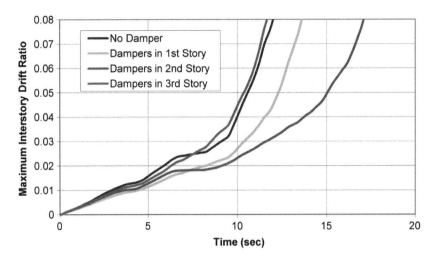

FIGURE 6.15 ET maximum interstory drift ratio curves for FM03B3RGW frames with different locations for dampers.

TABLE 6.6
Comparison between the results of nonlinear response history analysis for FM03B3RGW frame with different damper locations

| Records | No damper | Maximum interstory drift ratio | | |
		Damper in 1st story	Damper in 2nd story	Damper in 3rd story
LPAND270	0.0699	0.0401	0.0246	0.0534
LADSP000	N.A.	0.0219	0.0497	N.A.
MHG06090	0.0434	0.0534	0.0468	0.0456
LPGIL067	0.0244	0.0199	0.0250	0.0231
LPLOB000	0.0290	0.0241	0.0221	0.0287
NRORR360	N.A.	0.2797	0.0504	N.A.
LPSTG000	0.0444	0.0314	0.0335	0.0362

ET analysis and nonlinear response history analysis are conducted for this frame by installing the damper in different stories. The results of the analysis are shown in Figure 6.15 and Table 6.6. Figure 6.15 shows that by installing the damper in the third story, no significant performance improvement is achieved. The same result is also obtained by comparing the values of maximum interstory drift ratios for different records obtained from nonlinear response history analysis. Analysis does not converge for LADSP000 and NRORR360 records for both cases. Another point that can be concluded from Figure 6.15 is that, by installing the damper in the first or second stories, the performance of the frame improves. The best performance is obtained when the damper is installed in the second story. The performance in this case is really better in high IMs. Maximum interstory drift ratios of the frames having the dampers in the first or second story at the equivalent time are 0.0344 and 0.0272, respectively. Again, the results of nonlinear response history analysis confirm the results of ET analysis. Nonlinear response history analysis shows that the averages of maximum interstory drift ratios of the frames having the dampers in the first or second story are 0.0672 and 0.0360, respectively. The results of ET analysis somewhat underestimate the results of earthquakes, again, because of the P-Δ effects.

As mentioned previously, one of the beneficial advantages of the ET method is its capability in differentiating, with a minimum number of analyses, between different structural systems. The previous example shows that, although the results of ET analysis are not always exactly consistent with the results of ground motions analysis, the ET method can pinpoint the structure with better performance, even in the case of structures with relatively complicated behavior. This can be put into good use in performance-based seismic design where a lot of trial designs should be checked in order to find the structure that optimally meets the performance objectives.

6.6   SUMMARY AND CONCLUSIONS

In this chapter, application of the ET method in the analysis of multistory frames was demonstrated. Comparative case studies were provided to show how the analysis procedure works. In most frames with the EPP material model, estimations of ET analysis for maximum interstory drift ratio are less than nonlinear response history analysis results. The difference between the results is more in under-designed frames, which experience more nonlinearity in their analysis. But in frames that behave more linearly than others, the difference between the results of ground motions is less, and the results more closely match the results of ET analysis. The consistency of the base shears obtained by two methods is reasonable. The procedure to find the equivalent time in ET analysis to match the IMs of two methods is acceptable. The ET method is successful in locating the story with the maximum interstory drift ratio.

The dispersion of the results of nonlinear history analysis for the frames with the EPP material model that experience more nonlinearity is relatively high. The dispersion of results cannot be estimated using current sets of ET acceleration functions. When the dispersion of the results of nonlinear response history analysis is high, ET analysis more significantly underestimates the maximum interstory drift ratio.

The main reason for the dispersion of the results of nonlinear response history analysis is P-Δ effects. In cases where P-Δ effects are excluded, the results of two methods closely match. The same phenomenon can be seen in the results of ET analysis, but the effects of P-Δ on the results of ET analysis show up at a higher IM as compared to nonlinear response history analysis. Frames with the EPP material model are more sensitive to P-Δ effects because they develop a negative postyield stiffness under severe ground motions. Therefore, for some of the ground motions, P-Δ effects increase the maximum interstory drift ratio drastically, and the average value of this parameter is affected considerably, making it unusable. In STL material models that have 3 percent postyield stiffness, the results of ET analysis match non-linear response history analysis results with good precision.

It is shown that, although the results of ET analysis are not exactly consistent with the results of ground motions analysis in all cases of material properties and IMs, in most cases, the applied ET analysis is quite successful in differentiating between structures (or design alternatives), with better performance even in the case of relatively complicated structural responses. It should be noted that when ET analysis estimation for the EDPs of the frames shows almost identical results, some reservations should be considered before drawing conclusions. The differences observed in these cases may have resulted from purely random effects. A detailed nonlinear response history analysis, using a fairly large set of relevant earthquakes or improved ET acceleration functions, are needed for the final verification in cases where two systems show very similar results.

NOTE

1   *Chapter Source*: Riahi, H.T., H.E. Estekanchi, and A. Vafai. 2009. "Estimates of Average Inelastic Deformation Demands for Regular Steel Frames by the Endurance Time Method." *Scientia Iranica* 16, no. 5, pp. 388–402.

REFERENCES

American Institute of Steel. 1989. *Allowable Stress Design Manual of Steel Construction*, 9th ed. Chicago: AISC.
ASCE STANDARD, ASCE/SEI 7-05. 2006. *Minimum Design Loads for Buildings and Other Structures*. American Society of Civil Engineers.
ATC-40. 1996. *Seismic Evaluation and Retrofit of Concrete Buildings*. Report ATC-40. Redwood City, CA: Applied Technology Council.
BHRC. 2005. *Iranian Code of Practice for Seismic Resistant Design of Buildings*, 3rd ed. Standard No. 2800-05. Tehran: Building and Housing Research Center.
Chopra, A.K., R.K. Goel, and C. Chintanapakdee. 2003. "Statistics of Single-Degree-of-Freedom Estimate of Displacement for Pushover Analysis of Buildings." *Journal of Structural Engineering* 129, no. 4, pp. 459–469.
Estekanchi, H.E., K. Arjomandi, and A. Vafai. 2007. "Estimating Structural Damage of Steel Moment Frames by Endurance Time Method." *Journal of Constructional Steel Research* 64, no. 2, pp. 145–155.
Estekanchi, H.E., M. Mashayekhi, H. Vafai, G. Ahmadi, S.A. Mirfarhadi, and Harati, M. (2020). "A State-of-knowledge Review on the Endurance Time Method." *Structures* 27, pp. 2288–2299.
Estekanchi, H.E., A. Vafai, and M. Sadeghazar. 2004. "Endurance Time Method for Seismic Analysis and Design of Structures." *Scientia Iranica* 11, no. 4, pp. 361–370.
Estekanchi, H.E., V. Valamanesh, and A. Vafai. 2007. "Application of Endurance Time Method in Linear Seismic Analysis." *Engineering Structures* 29, no. 10, pp. 2551–2562.
Fajfar, P. 2000. "A Nonlinear Analysis Method for Performance-Based Seismic Design." *Earthquake Spectra* 16, no. 3, pp. 573–592.
FEMA-356. 2000. *Prestandard and Commentary for the Seismic Rehabilitation of Buildings*. Washington, DC: Federal Emergency Management Agency.
FEMA-440. 2005. *Improvement of Nonlinear Static Seismic Analysis Procedures*. Washington, DC: Federal Emergency Management Agency.

Foutch, D.A., and S. Shi. 1998. "Effects of Hysteresis Type on the Seismic Response of Buildings." Proceedings of 6th U.S. National Conference on Earthquake Engineering. Oakland, CA and Seattle, WA: Earthquake Engineering Research Institute.

Gupta, A., and H. Krawinkler. 2000. "Dynamic P-Delta Effects for Flexible Inelastic Steel Structures." *Journal of Structural Engineering* 126, no. 1, pp. 145–154.

Medina, R.A., and H. Krawinkler. 2005. "Evaluation of Drift Demands for the Seismic Performance Assessment of Frames." *Journal of Structural Engineering* 131, no. 7, pp. 1003–1013.

Miranda, E. 1999. "Approximate Seismic Lateral Deformation Demands in Multistory Buildings." *Journal of Structural Engineering* 125, no. 4, pp. 417–425.

OpenSees. 2002. *Open System for Earthquake Engineering Simulation.* Pacific Earthquake Engineering Research Center, http://peer.berkeley.edu/.

Riahi, H.T., and H.E. Estekanchi. 2006. "Application of Endurance Time Method for Estimating Maximum Deformation Demands of Structures." *First European Conference on Earthquake Engineering and Seismology.* Geneva.

Seneviratna, G.D.P.K., and H. Krawinkler. 1997. *Evaluation of Inelastic MDOF Effects for Seismic Design.* In John A. Blume, Earthquake Engineering Center Technical Report 120. Stanford Digital Repository, Available at: http://purl.stanford.edu/qt131wv6180.

Teran-Gilmore, A. 2004. "On the Use of Spectra to Establish Damage Control in Regular Frames." *Earthquake Spectra* 20, no. 3, pp. 995–1020.

Vamvatsikos, D., and C.A. Cornell. 2002. "Incremental Dynamic Analysis." *Earthquake Engineering and Structural Dynamics* 31, no. 3, pp. 491–514.

Whittaker, A., M. Constantinou, and P. Tsopelas. 1998. "Displacement Estimates for Performance-Based Seismic Design." *Journal of Structural Engineering* 124, no. 8, pp. 905–912.

7 Multicomponent Endurance Time Analysis

7.1 INTRODUCTION

Earthquake-induced ground motions have three translational components that can be recorded by accelerographs.[1] There are many situations that justify consideration of the effects of ground-motion components in the seismic analysis of sensitive structures. For example, three-dimensional analysis is recommended for asymmetric, tall buildings or important structures such as dams, bridges, and power plants (Wilson 2002). In these circumstances, the most appropriate analysis procedure is time-history analysis, including components of consistent ground motions. The endurance time (ET) method can readily be extended to multicomponent seismic analysis. Various methods and approaches can be adopted for extending ET analysis to three dimensions. Some ideas will be presented in this chapter. The reader is encouraged to follow the latest developments in this regard through state-of-the-art research in this area (Estekanchi et al. 2020).

With the development of new computational tools, the capability for realistic dynamic modeling and complex analysis of structures has increased and, in this situation, using improved and more complicated methods for seismic evaluation of structures has become a reasonable choice. Therefore, traditional two-dimensional static and response spectrum methods are gradually being replaced by nonlinear three-dimensional time-history analysis. In response to this increasing demand for application of these complex methods, it is necessary to develop procedures for clear and logical use of these new approaches.

Three-dimensional analysis using actual records has two major issues. First, for a particular site specification, the number of available recorded earthquakes might not be sufficient, and the selection of consistent accelerograms complicates the situation. Second, analysis of structures under these ground motions is very time consuming, especially when consideration of critical orientation is necessary. Moreover, interpretation of results for complex structures is quite difficult. Therefore, it is advantageous to use simpler methods that can estimate structural behavior under multidirectional excitation, with satisfactory approximation and less-computational demand.

The Endurance Time method is capable of being used in both the linear and nonlinear seismic analysis of structures (Estekanchi, Vafai, and Sadeghazar 2004). One of the advantages of this method over other time-history analysis procedures is in reducing the required computational effort, and its relative simplicity. In the ET method, the response of a structure is monitored against the intensity of excitation, from beginning to collapse – somewhat similar to the Incremental Dynamic Analysis method (Vamvatsikos and Cornell 2002). The structure is then assessed based on its response at various equivalent excitation levels.

In this chapter, the application of the ET method in multicomponent linear seismic analysis of structures is presented. The extension of the proposed procedure to more complicated situations should be clear. The ET method is evaluated by comparing results of the ET analysis with results of time-history analysis using horizontal components of real ground motions according to seismic analysis regulations, such as the Iranian National Building Code (INBC; Iranian Code of Practice for Seismic Resistant Design of Buildings) and ASCE 7-05 (ASCE 2006).

The first part of this chapter is devoted to a brief review of code regulations and some investigations on the three-dimensional analysis of buildings. In the next section, various structures designed according to the INBC code are analyzed by both ET and time-history analysis of the effects of real earthquakes. Finally, by comparing the results of these two methods, an algorithm for code-compliant ET analysis is proposed for simultaneous excitation in perpendicular directions of structures. Even though time-history analyses are seldom required in linear elastic analysis of structures, the current research is aimed at laying the foundations for extension of application of the ET method to seismic assessment using three-dimensional dynamic models subjected to realistic multicomponent ground motions. Obviously, major benefits of the procedure can only be realized when dealing with complicated nonlinear models. Even though nonlinear two-dimensional analysis results using currently available ET records indicate that reasonable estimates can also be obtained in nonlinear range, nonlinear multicomponent ET analysis is not covered in this chapter.

7.2 REVIEW OF CODE PROVISIONS AND RELATED RESEARCH

Although there are some guidelines in seismic codes for multidirectional analysis of the effects of real ground motions, these methods are not routinely applied in the seismic analysis and design of common buildings (Beyer and Bommer 2007). Considerable research has been conducted in the past to clarify and simplify three-dimensional analysis. Naeim, Alimoradi, and Pezeshk (2004) proposed the use of a genetic algorithm for selecting and scaling records. Many investigations have been performed to find characteristics of components of an earthquake (López 2006; Baker and Cornell 2006; Penzien and Watabe 1975) and the structural response due to two or three components of ground motions (Hernández and López 2002). These efforts led, not only to suggestions for the time-history analysis of structures, but also to recommendations for the application of components in static and response spectrum analysis (López, Chopra, and Hernández 2001; Zovani and Barrionuevo 2004).

Almost all structural design codes have basically the same recommendations for the selection of earthquake records for the purpose of three-dimensional analysis; however, they are somewhat different in the scaling method and application of components of records. For example, INBC, ASCE4-98, and EC8 recommend that analysis should be performed under components in principal directions of buildings (ASCE 2000; CEN 2003), but columns or walls intersecting seismic force-resistant systems of a building located in Categories E and F – as defined in the code, ASCE7-05 – necessitate application of ground-motions components in critical direction in addition to analysis using horizontal components along principal directions. FEMA (2001) recommends that each pair of time histories be applied simultaneously to the model, considering the most disadvantageous location of mass eccentricity.

One way to consider critical directions is by rotating the angle of induced excitation. With this procedure, an analysis requires a great deal of effort and time that might not be justified for typical structures. To avoid such problems, simplified methods have been proposed to estimate the critical response of structures to an earthquake without rotating the angle of excitation (Athanatopoulou 2005). However, rotating the angle of excitation is still more practical for considering the critical response of structures. The huge amount of computational effort required in three-dimensional response history analysis using bidirectional excitation at multiple levels can be prohibitive in many analysis and design situations. The ET method can considerably reduce the number of required analyses and, with appropriate approximation, provides a simple method for the three-dimensional analysis of structures. It should be noted again that, in this chapter, only linear behavior is investigated where GM analysis results at various excitation levels can be obtained by applying a scale factor, However, it should be obvious that this assumption cannot be used in general nonlinear cases.

7.3 ADAPTATION OF THE ENDURANCE TIME METHOD

7.3.1 THE BASIC CONCEPTS FROM THE ET METHOD

The ET method was introduced as a seismic analysis method, and application of this method in two-dimensional linear and nonlinear analyses of steel frames has been reported in the literature (Estekanchi, Valamanesh, and Vafai 2007; Riahi and Estekanchi 2010), as covered in Chapter 6. In the ET method, structures are subjected to a set of specially designed, intensifying accelerograms, called "ET acceleration functions" and their seismic performance is judged based on their response at various equivalent dynamic excitation intensities.

In ET analysis, simple interpretation of endurance time is considered to be the length of time required for the maximum value of the specified design parameter to exceed its allowable limit. In order to decide on whether the achieved performance can be considered to be adequate or not, the structural response at equivalent intensity of imposed dynamic action should be considered. Spectral acceleration is among the most popular intensity measures used in practice and has been considered for calibrating the ET acceleration functions used in this chapter.

Most ET acceleration functions are linearly intensifying over time. In this way, when the target time is set to $t = 10$ sec, it means that the considered ET acceleration functions are calibrated in such a way that their response spectra in a window from $t = 0$ to 10 sec match the design spectrum with a scale of unity. When the window of an acceleration function is taken from $t=0$ sec to $t=5$ sec, its response spectrum

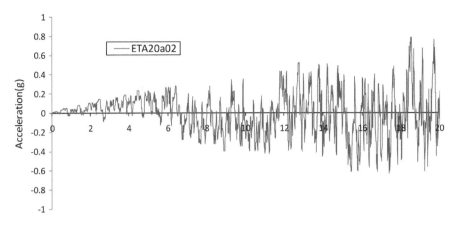

FIGURE 7.1   A typical ET acceleration function.

corresponds to half the template spectra at all periods and, if an interval of $t=0$ sec to $t=15$ sec is taken, its response spectrum matches with 1.5 times the template spectrum, and so on. Therefore, for a certain structure, which is designed according to a design spectrum that matches the template spectrum with a scale factor of unity at $t=t_{Target}$ (10 seconds in this research), and if, for example, the drift ratio exceeds its limit at $t=15$ sec, it can be concluded that the structure satisfies drift criteria, since its endurance time is more than what is required by the code – that is, a minimum time of 10 seconds, in this case. A typical ET acceleration function (second series from ETA20a series) is depicted in Figure 7.1.

In this chapter, the ETA20f series will be used for ET excitations. This set of ET excitations is not specifically generated to multicomponent analysis. For generation of ET acceleration functions used in this chapter, the concept of the response spectrum has been applied. By scaling the ET acceleration functions using a simple linear-scale factor, spectral acceleration and spectral displacement (Sa and Sd) can be set at any desired time to reach the required target level. By applying this method, we define the target response of ET acceleration functions as in Equations (7.1) and (7.2):

$$S_{aT}\left(T,t\right)=\frac{t}{t_{T \arg et}}S_{aC}\left(T\right) \tag{7.1}$$

$$S_{uT}\left(T,t\right)=\frac{t}{t_{T \arg et}}S_{aC}\left(T\right)\times\frac{T^{2}}{4\pi^{2}} \tag{7.2}$$

where $S_{aT}(T,t)$ is the target acceleration response at time t, T is the period of free vibration, $S_{aT}(T)$ is the codified design acceleration spectrum, and $S_{uT}(T,t)$ is the target displacement response at time t. The problem of generating accelerograms with such characteristics was approached by formulating it as an unconstrained optimization problem in the time domain, as follows:

$$\text{Minimize } F\left(a_g\right)= \int\limits_{0}^{T_{max}} \int\limits_{0}^{t_{max}} \left\{\left[S_a(T,t)-S_{aT}(T,t)\right]^2 + \alpha\left[S_u(T,t)-S_{uT}(T,t)\right]^2\right\}dt\,dT$$

(7.3)

where a_g is the ET accelerogram being sought and α is an optimization weighting parameter set to 1.0 in this study (Estekanchi, Valamanesh and Vafai 2007).

7.4　CHARACTERISTICS OF ET ACCELERATION FUNCTIONS USED IN THIS STUDY

Various sets of ET acceleration functions have been developed based on an intended application. In general, these fall into two major categories: code compliant and ground-motions compliant. Code compliant ET acceleration functions are based on a template spectrum that matches that of a particular design spectrum of a specified seismic code. These acceleration functions are mostly interesting from the design application perspective. On the other hand, ground-motions compliant ET records are based on the average response spectrum of a set of ground motions pertaining to specific soil conditions without any modifications to provide a safety margin. These records are more suitable when for comparative studies to analyze some inherent sources of inconsistency and scatter of the estimations obtained by the ET method. There are also some sets of ET excitation records, such as the ETA20in series, that are specifically generated to directly match different components of ground motions. In this chapter, however, a procedure will be explained which is not sensitive to ground-motion direction and thus can be more effective for practical engineering applications.

Major characteristics of ET acceleration functions that have the greatest influence on structural response match well with the ground motions (Valamanesh, Estekanchi, and Vafai 2010). This is mostly due to the fact that ET acceleration functions are designed in such a way as to produce response spectrums matching those of ground motions. In in this chapter, ETA20f01-03 acceleration functions, whose template spectrum matches with the average response spectrum of major components from seven real accelerograms (listed in FEMA 440 for soil type C), at the target time of a 10th of a second are used. Similar to other sets of ET acceleration functions, in this set the response spectra of these acceleration functions increase with time. In Figure 7.2, the response spectra of the ETA20f acceleration functions are compared at different times. As shown in Figure 7.2, linear intensification of response spectra at different times is apparent.

7.5　COMPARISON OF ET METHOD WITH CONVENTIONAL APPROACHES

In static analysis, by applying an equivalent load based only on a first mode shape, the effect of higher vibration modes of the structure is mostly excluded. By increasing the irregularities and complexities in buildings, the effects of dynamic specifications become remarkable, and static analysis will not be reliable. The ET method is based

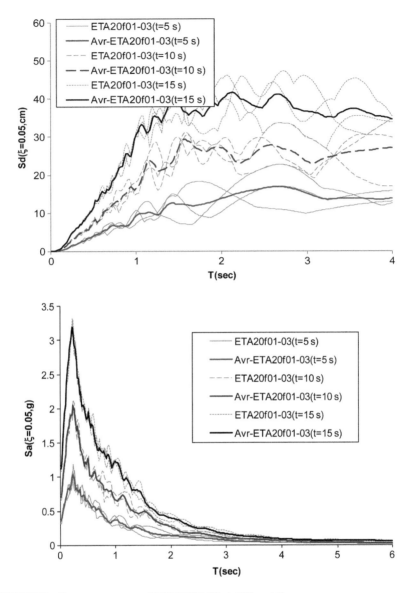

FIGURE 7.2   Response spectra of ETA20f01-03 at different times.

on time-history analysis and can intrinsically take all significant dynamic properties of the structure into account. Moreover, because ET acceleration functions are
intensifying with time, in each ET analysis the performance of the structure can be
predicted at different levels of intensity, while the analysis of a building with ground
motions at different levels needs Incremental Dynamic Analysis (IDA) (Vamvatsikos
and Cornell 2002), which requires a considerable computational effort. This advantage of the ET method cannot be fully realized in linear analysis, which is the subject

of this chapter. While some particular problems, such as optimal damper placement in linear systems, still require a response history-based analysis procedure, the major goal in this study should be considered as proposing some methods and ideas in order to show a way to extend the application of ET into multicomponent seismic analysis.

7.6   STRUCTURAL MODELS

In the model studied here, the endeavor is to focus on those parameters that are more significant in three-dimensional analysis. Several steel moment-resistant frames with one, three, four, five, and seven stories in three states – regular, irregular in one direction, and irregular in two directions – for considering the effects of torsion, are designed and investigated. It should be noted that for all frames, the story height is assumed to be 3.2 m and all spans are equally 6 m. Box sections were assigned to columns, and HE-A (European wide flange I section) profiles were assigned to beams.

Models are named based on their lateral load-resisting system, the number of stories, spans in both directions, and irregularities in each direction, as follows: All frame names begin with F3DMM, signifying that all of them are 3D moment frames in both directions. This is followed by the letter S and a number that shows the number of stories. Then the number of bays in X and Y directions is specified as XnYm meaning n bays in the X direction and m bays in the Y direction. Irregularity of the frame in X or Y or both directions is indicated next – for example, IRX means irregular in X direction and so on. For example, as shown in Figure 7.3b, F3DS3X3Y3IRXY represents a three-story moment-resistant building, with three spans in X, Y directions, and irregularities in both directions.

The equivalent static lateral force procedure, based on the provisions of INBC (INBC 2005) for soil type 2, has been used for the design of frames. Dead and live loads are assumed at 7,500 and 2,500 N/m2, respectively, and an accidental eccentricity of 0.05L (where L is dimension of the plan of the building in each direction) is considered for the design per code's requirements. The damping ratio for all frames is assumed to be 0.05, a typical value for this type of structure. Beam and column profiles are HE-A (wide flange) and Box profiles, respectively. The importance factor is assigned to be 1 and the R factor is considered to be 7 in both directions due to the moment-resistant frame in both directions. Properties of frames and design assumptions are listed in Table 7.1. These buildings have predominant periods between 0.1 and 1.5 seconds. It seems that by covering a reasonable range of model variety, the results of ET analysis can be extended for three-dimensional analysis of low-rise steel moment frames.

7.7   SELECTION OF REFERENCE GROUND MOTIONS

To verify results of the ET method with ground motions, seven real accelerograms are selected from 20 records listed in FEMA 440 for Soil Condition C. These records and their components are listed in Table 7.2. In this chapter, the effect of a vertical component is not included. The average response spectra of these real accelerograms, which are scaled according to code requirements, are illustrated in Figure 7.4.

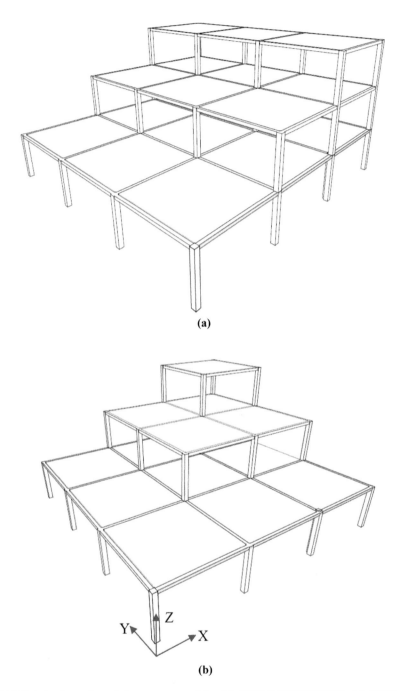

(a)

(b)

FIGURE 7.3 Investigated three-story models, (a) F3DMMS3X3Y3IRX, (b) F3DMMS3 X3Y3IRXY.

TABLE 7.1
Investigated Frames, Properties and Design Assumptions

Name	No. stories	No. span X	No. span Y	Seismic coefficient	T(sec)	Base shear (KN)
F3DMMS3X3Y3IRX	3	3	3	0.125	0.722	735
F3DMMS3X3Y3IRXY	3	3	3	0.125	0.627	601
F3DMMS4X3Y3	4	3	3	0.119	0.93	1,411
F3DMMS4X3Y3IRX	4	3	3	0.119	0.913	1,270
F3DMMS4X3Y3IRXY	4	3	3	0.119	0.875	810
F3DMMS5X4Y4	5	4	4	0.106	1.08	2,826
F3DMMS5X4Y4IRX	5	4	4	0.106	0.996	2,209
F3DMMS5X4Y4IRXY	5	4	4	0.106	0.976	2,143
F3DMMS7X3Y5	7	3	5	0.091	1.505	3,159

TABLE 7.2
Properties of Real Accelerograms and Their Components

Name	Ms	Station name	Abbreviation	Component (deg)	PGA (cm/s^2)
Landers	7.5	Yermo, Fire Station	LADSP000	0	167.80
			LADSP090	90	151.05
Loma Prieta	7.1	Saratoga, Aloha Ave.	LPSTG000	0	494.50
			LPSTG090	90	317.90
Loma Prieta	7.1	Gilroy, Gavilon College Phys Sch Bldg	LPGIL067	67	349.10
			LPGIL337	337	318.80
Loma Prieta	7.1	Santa Cruz, University of California	LPLOB000	0	433.10
			LPLOB090	90	387.00
Loma Prieta	7.1	Anderson Dam, Downstream	LPAND270	270	239.40
			LPAND360	360	235.10
Morgan Hill	6.1	Gilroy #6, San Ysidro Microwave Site	MHG06090	90	280.40
			MHG06000	0	217.87
Northridge	6.8	Castaic, Old Ridge Route	NRORR360	360	504.20
			NRORR090	90	557.30

One important point in Figure 7.4 is in the difference between the response spectra of horizontal components of each ground motion. Although the spectrum of each component is not the same at different periods, especially between 0.5 and 3 seconds, for the general purpose of seismic analysis in this study, this difference is assumed to be insignificant due to the fact that each record is applied in orthogonal directions, thus maximum response is assumed to be the significant parameter anyway. In the ET analyses in this chapter, ET acceleration functions with the same intensity and spectral shape are used in the bidirectional analysis of studied frames.

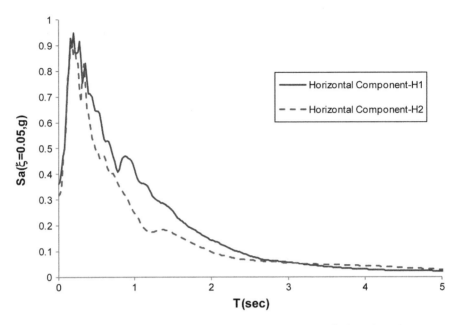

FIGURE 7.4   Average response spectra of horizontal components of selected accelerograms.

7.8   MULTICOMPONENT ANALYSIS

7.8.1   SCALING PROCEDURE

There are different approaches for scaling earthquake records, such as the square root of the sum of the squares (SRSS), arithmetic and geometric mean and the maximum spectral response. Among many types of averaging spectral components, in this research SRSS is selected because of its relatively better fit with the target spectrum. According to ASCE 7-05, horizontal components of ground motions shall be scaled in such a way that the average SRSS spectrum from all horizontal component pairs, in range of 0.2T to 1.5T, where T is the predominant period of vibration for studied structure, does not fall below 1.3 times of corresponding ordinate of design spectrum by more than 10 percent. This approach is used for scaling the components of ground motions. These scaling values for used ground motions are illustrated in Table 7.3.

The scaling procedure for applying ET acceleration functions resembles scaling of actual records, that is, mentioned methods are used to obtain the scale factor for ET acceleration functions considering their response spectrum at target time. For example, for a pair of ET acceleration functions that consist of ETA20f01 and ETA20f02, the acceleration response spectrum for each ETAF (Endurance Time Excitation Function) is calculated at target time. Then using the SRSS method, mentioned above, these response spectra are combined and compared to the amplified design spectrum (the design spectrum multiplied by a factor of 1.3), the scaling factor could be calculated, which should be applied for both used ET acceleration functions. In this way, not

TABLE 7.3
Scaling Value of Records Components Used in Analysis of Frames

	LADSP	LPSTG	LPGIL	LPLOB	LPAND	MHG06	NRORR
F3DMMS3X3Y3IRXY	1.805	0.988	1.042	1.017	1.158	1.078	0.533
F3DMMS3X3Y3IRX	1.734	0.986	1.097	1.082	1.203	1.062	0.527
F3DMMS4X3Y3	1.678	0.954	1.192	1.246	1.276	1.024	0.526
F3DMMS4X3Y3IRX	1.678	0.954	1.192	1.246	1.276	1.024	0.526
F3DMMS4X3Y3IRXY	1.678	0.954	1.192	1.246	1.276	1.024	0.526
F3DMMS5X4Y4	1.655	0.925	1.228	1.329	1.292	1.036	0.523
F3DMMS5X4Y4IRX	1.655	0.925	1.228	1.329	1.292	1.036	0.523
F3DMMS5X4Y4IRXY	1.655	0.925	1.228	1.329	1.292	1.036	0.523
F3DMMS7X3Y5	1.555	0.842	1.314	1.599	1.304	1.121	0.515

TABLE 7.4
Scaling Value for Pairs of ET Acceleration Functions

	ETA20f01,02	ETA20f02,03	ETA20f03,01	Average ET
F3DMMS3X3Y3IRXY	0.473	0.475	0.478	0.475
F3DMMS3X3Y3IRX	0.482	0.482	0.483	0.482
F3DMMS4X3Y3	0.480	0.483	0.483	0.482
F3DMMS4X3Y3IRX	0.480	0.483	0.483	0.482
F3DMMS4X3Y3IRXY	0.480	0.483	0.483	0.482
F3DMMS5X4Y4	0.485	0.488	0.484	0.486
F3DMMS5X4Y4IRX	0.485	0.488	0.484	0.486
F3DMMS5X4Y4IRXY	0.485	0.488	0.484	0.486
F3DMMS7X3Y5	0.495	0.499	0.497	0.497

only did the results from all scaling approaches lead to almost the same factor for ET acceleration functions, but this scaling factor did not change significantly from one frame to another; while these scaling factors were considerably different in various ground motions due to their specific response spectrum. The major reason for such consistency of scaling methods in ET acceleration functions is that they inherently comply with the design response spectrum and, so, the shape of the response spectrum will be almost the same in different accelerations functions belonging to the same set of records. The scale factors for pairs of ET acceleration functions are shown in Table 7.4. As shown in Table 7.4, the average scale factor of three pairs of ET acceleration function is used for all individual pairs in the analysis of each model.

7.8.2 MULTICOMPONENT ANALYSIS BY THE ET METHOD

ET acceleration functions used in this study are designed in such a way that their response spectra increase linearly over time. When used for response history analysis, most of the regulations set forth in design codes, regarding the general three-dimensional time-history analysis, are also applicable for ET analysis. However, some special characteristics of ET acceleration functions require particular consideration. Although ET acceleration functions are statistically independent, all ET acceleration functions in the same set are produced in the same manner and use the same assumptions; thus, statistically, the intensity and response spectrum at each time are, theoretically, the same for all ET acceleration functions in a set of ET acceleration functions. Therefore, the definition of a major or a minor component in these types of ET records is not relevant. Second, as the ET acceleration functions are produced synthetically, critical angle or principal direction of excitation are also of little significance. Finally, when all ET acceleration functions in the ET excitation set are statistically alike, pairs of ET acceleration functions can be considered by swapping ET acceleration functions alternatively with each other; that is, the first pair of ET excitations include ETA20f01 in X direction and ETA20f02 in Y direction; the second pair is a combination of ETA20f02 in X direction and ETA20f03 in Y direction; and the third pair is made up of ETA20f03 and ETA20f01 in X and Y directions, respectively. These pairs are applied to the structure alternately, and results are averaged for final evaluation. It should be noted that this approach is one among many different approaches that can be adopted for the purpose of multicomponent ET analysis. The interested reader is encouraged to also explore other approaches and procedures in this regard. A proposed algorithm for three-dimensional ET analysis used here is illustrated in Figure 7.5.

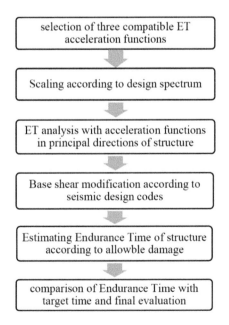

FIGURE 7.5   Proposed flowchart for bidirectional analysis of structures by ET method.

Following the flowchart in Figure 7.5, the designed frames were analyzed and compared with results from time-history analysis under previously mentioned real accelerograms in a situation in which components of records are applied in principal directions of structures. For instance, displacements in X and Y directions of a three-story building, obtained from two methods, are compared.

As in ET analysis, time is a representative of intensity; it is obvious that results from the ET method are plotted over the time, while responses of real accelerograms appear as points with +/− one standard deviation mark and are extended by a line (representing the linear analysis) for comparison. These values are compared with the ET method at target time, that is, $t=10$ sec in this study (Figure 7.6). In this figure, Uxsti and Uysti determine ith story displacement in X and Y directions, respectively. As shown, the results of ET analysis at $t=10$ sec are close to the results obtained from analysis under real accelerograms in principal directions. It should be noticed that the curve is an average of results from ET analysis and points are the average of results from real accelerograms. Further investigations show that other frames had similar results. For example, drifts of a seven-story building in X direction obtained from ET analysis at $t=10$ sec are compared with results from real accelerograms (Figure 7.7).

In addition to displacement and drifts, internal forces of all members – for example, moments in beams and columns, and axial force in columns, are studied. In Figure 7.8, for a three-story building, moments and axial forces in some random beams and columns are sketched by time for ET analysis and compared with real accelerograms. In this figure, M_Bi and P_Cj refer to the maximum moment in beam number i, and maximum axial force in Column Number j, respectively. These elements are specified in Figure 7.3b. It is obvious from Figure 7.8 that the response of all studied structural indices in studied frames are approximately the same as in the ET method at target time ($t=10$ sec) and horizontal components of real ground motions in principal directions. Obviously, there are some discrepancies that will be discussed later in this chapter.

In addition to detailed study of some randomly selected members, all members, including all beams and columns, were investigated to specify any member that might show significantly different behavior from others. Furthermore, in this step the correlation of results between the ET method and ground motions is derived. In Figure 7.9, drifts and displacements of stories in both directions for the average of ground motions are drawn versus the average of ET analysis at the target time for the regular five-story frame. In addition, in Figure 7.10, the same figures are shown for the maximum moment in beams and axial force in all columns of the five-story building, which is irregular in both directions.

It is essential to note that the response of these structures is compared only under lateral loads and, in this state, the effect of vertical loads, such as gravity load and the effect of vertical acceleration, is not included. It should be clear that inclusion of the gravitational loads does not affect the conclusions obtained in this chapter, which are based on lateral load response.

As indicated from the figures, for studied damage criteria, the correlation of results from the ET method and real earthquakes is close to 1, and results from the average of earthquakes in principal directions can be estimated by a unique correction factor for

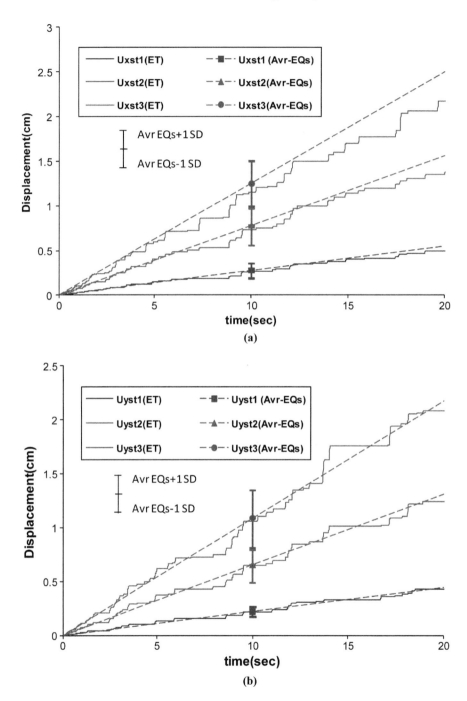

FIGURE 7.6 Displacement responses at any time in ET analysis and comparison with F3DMMS3X3Y3IRX, (a) displacement in X direction, (b) displacement in Y direction.

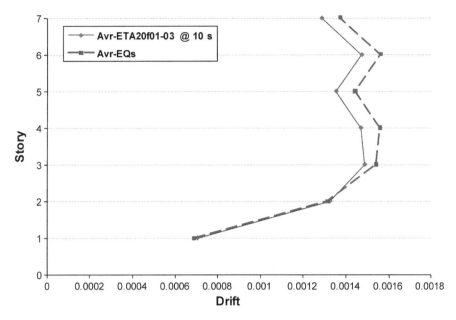

FIGURE 7.7 Comparison of drifts from ET analysis at t=10 sec and actual records, F3DMMS7X5Y3.

each frame. The correction factor is defined as the relation between results from real earthquakes and ET analysis at target time (t=10 sec), that is,

$$CF = \frac{DI_{Avr-EQs}}{DI_{ET@t=10}} \tag{7.4}$$

Correction factors and correlation coefficients of all studied frames and most damage criteria for each frame are shown in Table 7.5.

As can be seen in Table 7.5, the correlation coefficient between the ET method and ground motions for various responses of studied frames is near unity. This means that all members conform to a single correction factor (CF) and with application of this factor, the average response of real ground motions in principal directions can be estimated with reasonable accuracy by the ET method. The next point in this table is that the correction factors are nearly the same for various response parameters in each frame; thus, there is no need to apply different CF for different parameters. It is also found that the discussed CFs for all frames are about unity (with s maximum of 15 percent tolerance), meaning that results from the ET method at target time are the same as results from the average of real accelerograms in the principal directions of the structure.

As can be seen, there are some differences between results of ET acceleration functions and ground motions, these differences mostly occur due to the incompatibilities of response spectra of applied ET acceleration functions and actual

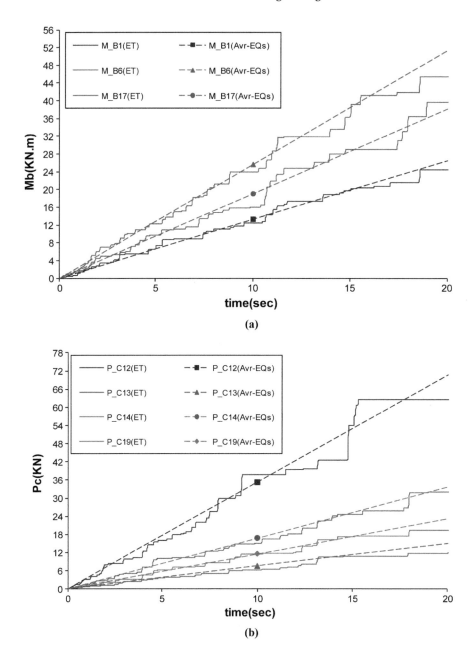

FIGURE 7.8 Internal force in members of F3DMMS3X3Y3IRXY in ET method and real earthquakes, (a) moment in beams, (b) axial force in columns.

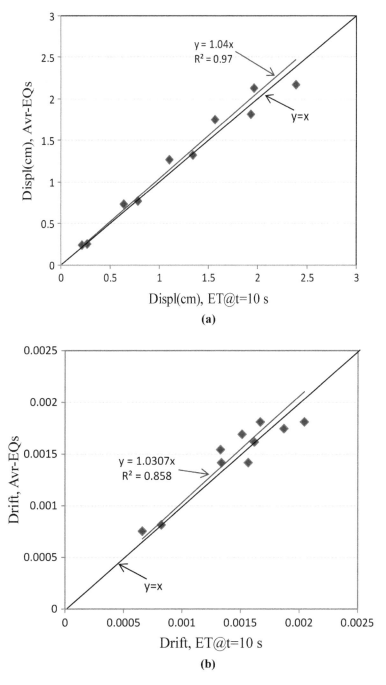

FIGURE 7.9 Drifts and displacement values from earthquakes versus ET results at t=10 sec for F3DMMS5X4Y4, (a) displacement, (b) drift.

TABLE 7.5
Correction Factor and Correlation Coefficient of Structural Responses in ET Method and Real Earthquake

	Displ		Drift		Mb		Pc		MxC		MyC	
	CF	R	CF	R	CF	R	CF	R	CF	R	CF	R
F3DMMS3X3Y3IRX	1.04	1.00	1.03	0.99	1.02	0.95	1.02	0.99	1.03	0.90	0.97	0.99
F3DMMS3X3Y3IRXY	1.00	1.00	0.95	0.99	0.99	0.96	0.96	0.99	0.97	0.93	0.99	0.84
F3DMMS4X3Y3	1.04	0.99	1.03	0.98	1.08	0.99	1.09	1.00	1.08	0.98	1.09	0.97
F3DMMS4X3Y3IRX	1.06	0.99	1.06	0.96	1.12	1.00	1.11	0.99	1.16	1.00	1.15	1.00
F3DMMS4X3Y3IRXY	1.05	0.99	1.05	0.98	1.12	1.00	1.11	0.99	1.12	1.00	1.11	1.00
F3DMMS5X4Y4	1.04	0.99	1.03	0.94	1.13	0.99	0.99	0.98	1.16	1.00	1.15	0.99
F3DMMS5X4Y4IRX	0.99	1.00	0.98	0.99	1.00	0.99	0.97	1.00	0.99	0.99	1.08	1.00
F3DMMS5X4Y4IRXY	1.01	0.98	0.99	0.96	1.04	0.91	1.04	0.99	0.98	0.99	1.15	1.00
F3DMMS7X5Y3	1.09	0.98	1.07	0.92	1.15	0.99	1.13	0.99	1.16	1.00	1.11	1.00

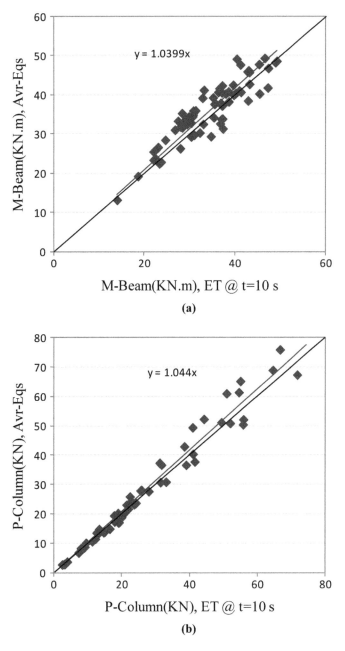

FIGURE 7.10   Internal forces from earthquakes versus ET results at t=10 sec for F3DMM S5X4Y4IRXY, (a) Moment in beams, (b) Axial force in columns.

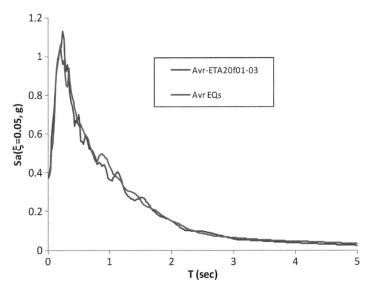

FIGURE 7.11 Average response spectrum of ETA20f01-03 at t=10 sec and average of maximum response of two horizontal components of real earthquakes.

ground motions. As can be seen in Figure 7.11 the average response spectrum obtained from the maximum response of two horizontal components for seven selected earthquake ground motions is not exactly the same, and at most periods of vibrations is a bit greater than the average response spectrum of three ET acceleration functions This inconsistency happens when, at some periods of vibrations, the response spectrum of the second component of each ground motion is greater than the first component, which is compatible with the produced ET acceleration functions. Also, the discrepancy is caused by the roughness of the target spectrum and optimization problems in generating ET acceleration functions.

To reduce these discrepancies, the compatibility between two spectra should be improved. This goal can be achieved by producing improved ET acceleration functions or using more than three acceleration functions in ET analysis. Also, instead of considering the first component of earthquake ground motions, the maximum response of two horizontal components can be considered for generating or scaling of ET acceleration functions. However, due to the fact that the ratio of intensities for two horizontal components is not well known, and there is no unique value for such parameter, it could be assumed that the ET acceleration functions are produced to be compatible with the component that has greater intensity. Because of this assumption, depending on the structural period, the results from the ET method may involve a slight underestimation as compared to those obtained from actual ground motions. Due to the fact that this incompatibility can be ignored in the current study considering the insignificance of the differences (maximum difference is 20 percent), this set of ET acceleration functions can be regarded as acceptable for a reasonable response estimation.

As stated at the beginning of this chapter, most seismic codes accept the application of seismic excitation only in principal directions; however, some structural codes, such as ASCE7-05, impose more stringent requirements, such as obligating that the analysis of members in intersections of two lateral-resistant systems of buildings – located in E and F seismic category – be performed in the critical direction. According to this requirement, engineers should analyze the structure using components of each earthquake at its critical direction. Then, maximum values obtained from each record are averaged from seven accelerograms. Although it is not likely that all members reach their maximum value simultaneously in the critical direction, and this approach seems to be conservative, it may be necessary to perform this type of analysis for important or critical structures. In this respect, the average of maximum structural response in the most adverse direction will be evaluated next. However, more investigation is required to draw general conclusions in this regard. In Figure 7.12, internal forces of all members for an irregular three-story frame are compared between an average of maximum results of each earthquake at their critical direction and the ET method at target time.

It is apparent that correlations of results from these two methods are satisfactory, and a correction factor can be applied to estimate the average of maximum results of earthquakes by the ET method. For studied frames, these correction factors are obtained and shown in Table 7.6. It should be noted that while strong correlation exists in each case, the correction factor varies based on the model, and no clear trend can be observed in order to propose a generally applicable correction factor.

Obviously, from Table 7.6, correlations for all frames and all damage criteria are significantly high and, for each frame, results from the ET method could be scaled up to results from the average of real accelerograms at their critical angles. One reason for this is that, when maximum responses of earthquakes at a critical angle for each ground motion are averaged, the effective level of response spectrum as an index of intensity increases as a result of the statistical process of maximizing between more analysis cases. On the other hand, the probability of exceedance of seismic hazard is reduced (Bazzurro et al. 1998). For example, the average of the maximum response in X direction of a 2DOF system under components of considered earthquakes at their critical directions were computed and compared with that of ET acceleration functions at target time (t=10 sec) in Figure 7.13. It is obvious that ET acceleration functions are applied just at two orthogonal directions and will not be rotated, and the critical angle for ET analysis in this case is not meaningful. It is seen that the response spectrum of ET acceleration functions is less than the response of the 2DOF system under horizontal components of real earthquakes at their critical angles at most periods of vibration. More studies are required in order to obtain effective response spectra pertaining to ground motions applied at all directions. In this way, ET acceleration functions can be developed based on these critical direction spectra and improved estimates can be made. However, it is also possible to improve the estimation by upscaling current ET records so that their response in Figure 7.13 matched those of ground motions at their critical angle. The spectral ratio of horizontal components of earthquakes at a critical direction and those at a principal direction and ET acceleration functions are depicted in Figure 7.14.

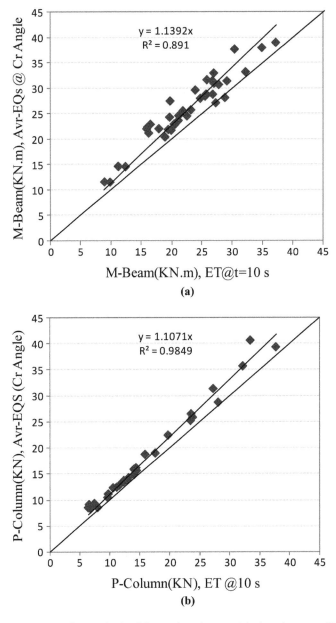

FIGURE 7.12   Internal forces obtained for earthquakes at critical angle versus ET method at target time F3DMMS3X3Y3IRXY, (a) moment in beams, (b) axial force in columns.

As can be seen in Figure 7.14, the spectral ratio of horizontal components for actual ground motions vary between 1 and 1.2 and for ETAFs are between 0.9 and 1.4. Furthermore, it is seen that at the periods $T>3$ sec, the ratio of the spectrum from earthquakes at the critical angle and ET acceleration functions increase. This

TABLE 7.6
Correction Factor and Correlation Coefficient of Structural Responses in ET Method and Real Earthquake at Their Critical Directions

	Displ		Drift		Mb		Pc		MxC		MyC	
	CF	R	CF	R	CF	R	CF	R	CF	R	CF	R
F3DMMS3X3Y3IRX	1.16	1.00	1.15	0.99	1.16	0.95	1.16	0.99	1.18	0.93	1.09	0.99
F3DMMS3X3Y3IRXY	1.12	1.00	1.07	0.99	1.14	0.96	1.11	0.99	1.12	0.92	1.14	0.76
F3DMMS4X3Y3	1.16	0.99	1.15	0.96	1.22	1.00	1.17	0.99	1.21	0.98	1.22	0.97
F3DMMS4X3Y3IRX	1.16	0.98	1.16	0.95	1.14	1.00	1.20	0.99	1.17	1.00	1.17	1.00
F3DMMS4X3Y3IRXY	1.19	0.98	1.19	0.94	1.27	1.00	1.20	0.99	1.17	1.00	1.17	1.00
F3DMMS5X4Y4	1.09	0.98	1.08	0.92	1.20	0.99	1.08	0.99	1.16	1.00	1.14	0.99
F3DMMS5X4Y4IRX	1.18	1.00	1.18	0.99	1.21	0.99	1.22	0.99	1.19	0.99	1.23	1.00
F3DMMS5X4Y4IRXY	1.16	0.98	1.12	0.96	1.18	0.90	1.23	0.99	1.17	0.99	1.22	1.00
F3DMMSX5Y3	1.01	0.98	1.00	0.90	1.09	0.99	1.08	0.99	1.10	1.00	1.05	1.00

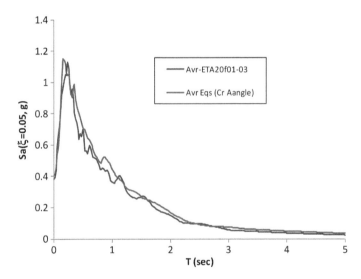

FIGURE 7.13 Average acceleration response of 2DOF system in x-direction under components of real earthquakes at their critical orientation and ETAf01-03at t=10 sec.

observation could be expected from Figure 7.4. The response spectrum from the second component of earthquakes at higher periods, that is, after T=3 sec, is greater than that of the first component, with which ET acceleration functions are consistent. Thus, at higher periods, special consideration should be given for determining the design spectrum, based on which the ET acceleration functions are selected or produced. Also, it is seen that the curve obtained for actual ground motions is smoother than that of ETAFs. It is due to the fact that the response spectra of ETAFs and used records are not exactly the same, and there are always minor differences between ETAFs

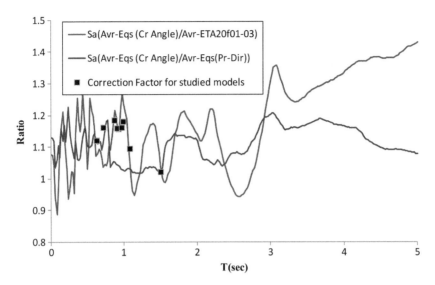

FIGURE 7.14   The comparison of spectral ratio of horizontal components of earthquakes at critical direction and principal direction and ET acceleration functions.

and the target spectrum. By the way, this figure is consistent with results obtained from Table 7.6, where the scale factor varies between 1 and 1.23. Comparing the CF obtained from Table 7.6 with Figure 7.14, it is concluded that the differences from ET analysis and results of time-history analysis at the critical angle can be interpreted by their response spectra, It seems that an appropriate scale factor, estimated from Figure 7.14, can be applied to studied frames to estimate the average response of real earthquakes at their critical angle. It should also be noted that this required statistical correction factor is conceptually the same, considering either ground motions or ET analysis results. On the other hand, in ET analysis, this scale factor can be converted into its equivalent extra time and thus, the average response of earthquakes at a critical angle can be estimated in the ET method by reading the response at a higher analysis time, provided that this observation can be verified considering more elaborate studies. Anyway, it should be noted that the response at critical direction can be quite different from that obtained from analysis based on orthogonal direction excitation, and further research in this area is required in order to reach general conclusions.

7.9   SUMMARY AND CONCLUSIONS

In this chapter, application of the ET method in the analysis of buildings under bidirectional excitation was investigated, and a procedure for three-dimensional analysis by the ET method was proposed. Following seismic code recommendations, results of the time-history analysis of buildings under horizontal components of earthquakes at principal directions and critical directions were compared with the results of ET analysis under pairs of ET acceleration functions applied in principal directions of the studied buildings. The following conclusions can be drawn:

1. The response of structures estimated by proposed bidirectional ET analysis procedure matches well with results from time-history analysis using real earthquake components in principal directions of structures.
2. The average and minimum correlation coefficient for analysis results obtained from the ET method and time-history analysis using real earthquakes for investigated frames are 0.97 and 0.80, respectively. Considering this strong correlation between the results, it can be concluded that the average response to seismic excitation in linear range can with reasonable accuracy be predicted by the ET method.
3. Response of structures studied in this chapter at critical directions of each earthquake can be correlated to their response using orthogonal direction analysis by applying a correction factor of about 1.05 to 1.2 in studied models. In these cases, results from ET analysis, at t=10 sec, can be multiplied by a correction factor or, alternatively, damage values should be read as a higher target time on the ET response curve for critical-direction estimations. However, this observation cannot be extended to general cases, and more investigation is required before a reasonable conclusion can be reached in this regard.
4. Based on the results from the studied models in this chapter, the response of steel moment frames subjected to multicomponent seismic excitation can be predicted with reasonable accuracy by using the proposed procedure. This procedure can reduce the required computational effort when time-history analysis is required, such as the analysis of the effect of damping devices. In this chapter, however, linear elastic models were considered in order to take full advantage of the ET method in multicomponent seismic analysis – extension of these concepts to nonlinear analysis should also be considered.

7.10   NOMENCLATURE

$a_g(t)$	Ground acceleration
ET	Endurance Time
ETacc	Endurance Time acceleration function
T	Free vibration period (s)
S_a	Acceleration response
$S_a(T,t)$	Acceleration response for period T at time t
$S_{aC}(T)$	Codified design acceleration spectrum for period T
$S_{aT}(T,t)$	Target acceleration response for period T at time t
$S_u(T,t)$	Displacement response for period T at time t
$S_{uT}(T,t)$	Target displacement response value for period T at time t
t	Time
t_{max}	Time corresponding to the end of accelerogram
T_{max}	Maximum free vibration period(s) to be considered in the optimization

$t_{T\arg et}$	Target time
α	Weighing factor in optimization target function
MMF	Moment-resistant frame in both directions
CF	Correction factor
$DI_{Avr-EQs}$	Damage Index obtained from averaging the response of earthquakes
$DI_{ET@t=10}$	Damage Index obtained under ET acceleration functions at t=10 sec.
ρ	Correlation Coefficient
Mb	Moment in Beams
Pc	Axial Force in Columns
MxC	Moment at the End of Column in X direction
MyC	Moment at the End of Column in Y direction
2DOF	Two-degrees-of-freedom system

NOTE

1   *Chapter Source*: Valamanesh, V., and H.E. Estekanchi. 2011. "Endurance Time Method for Multi-Component Analysis of Steel Elastic Moment Frames." *Scientia Iranica* 18, no. 2, pp. 139–142.

REFERENCES

ASCE. 2000. *Seismic Analysis of Safety-Related Nuclear Structures and Commentary*. ASCE Standard No. 004-98, American Society of Civil Engineers.

ASCE. 2006. *Minimum Design Loads for Buildings and other Structures*. ASCE Standard No. 007–05, American Society of Civil Engineers.

Athanatopoulou, A.M. 2005. "Critical Orientation of Three Correlated Seismic Components." *Engineering Structures* 27, pp. 301–312.

Baker, J.W., and C.A. Cornell. 2006. "Correlation of Response Spectral Values for Multicomponent Ground Motions." *Bulletin of the Seismological Society of America* 96, no. 1, pp. 215–227.

Bazzurro, P., C.A. Cornell, N. Shome, and J.E. Carballo. 1998. "Three Proposals for Characterizing MDoF Non-Linear Seismic Response." *Journal of Structural Engineering* 124, no. 11, pp. 1281–1289.

Beyer, K., and J.J. Bommer. 2007. "Selection and Scaling of Real Accelerograms for Bi-Directional Loading: A Review of Current Practice and Code Provisions." *Journal of Earthquake Engineering* 11, no. 1, pp. 13–45.

CEN. 2003. *Eurocode 8: Design of Structures for Earthquake Resistance. Part 1: General Rules, Seismic Actions and Rules for Buildings*. Brussels: European Committee for Standardization.

Estekanchi, H.E., M. Mashayekhi, H. Vafai, G. Ahmadi, S.A. Mirfarhadi, and M. Harati, (2020). "A State-of-knowledge Review on the Endurance Time Method." *Structures* v27, pp. 2288–2299.

Estekanchi, H.E., A. Vafai, and M. Sadeghazar. 2004. "Endurance Time Method for Seismic Analysis and Design of Structures." *Scientia Iranica* 11, no. 4, pp. 361–370.

Estekanchi, H.E., V. Valamanesh, and A. Vafai. 2007. "Application of Endurance Time Method in Linear Seismic Analysis." *Engineering Structures* 29, no. 10, pp. 2551–2562.

FEMA. 2001. *NEHRP Recommended Provisions for Seismic Regulations for New Buildings and other Structures*, 2000 Edition, *Part 1: Provisions*. FEMA 368, Washington, DC: Building Seismic Safety Council for the Federal Emergency Management Agency.

Hernández, J.J., and O.A. López. 2002. "Response to Three-Component Seismic Motion of Arbitrary Direction." *Earthquake Engineering and Structural Dynamics* 31, no. 1, pp. 55–77.

INBC. 2005. *Iranian Code of Practice for Seismic Resistant Design of Buildings*, Standard No. 2800-05, 3rd ed, Building and Housing Research Center, Tehran.

López, O.A. 2006. "Response Spectra for Multicomponent Structural Analysis." *Earthquake Spectra* 22, no. 1, pp. 85–113.

López, O.A., A.K. Chopra, and J.J. Hernández. 2001. "Evaluation of Combination Rules for Maximum Response Calculation in Multicomponent Seismic Analysis." *Earthquake Engineering and Structural Dynamics* 30, no. 9, pp. 1379–1398.

Naeim, F., A. Alimoradi, and S. Pezeshk. 2004. "Selection and Scaling of Ground Motion Time Histories for Structural Design Using Genetic Algorithms." *Earthquake Spectra* 20, no. 2, pp. 413–426.

Penzien, J., and M. Watabe. 1975. "Characteristics of 3-Dimensional Earthquake Ground Motion." *Earthquake Engineering and Structural Dynamics* 3, no. 4, pp. 365–373.

Riahi, H.T., and H.E. Estekanchi. 2010. "Seismic Assessment of Steel Frames with Endurance Time Method." *Journal of Constructional Steel Research* 66, no. 6, pp. 780–792.

Valamanesh, V., H.E. Estekanchi, and A. Vafai. 2010. "Characteristics of Second Generation Endurance Time Accelerograms." *Scientia Iranica* 17, no. 1, pp. 53–61.

Vamvatsikos, D., and C.A. Cornell. 2002. "Incremental Dynamic Analysis." *Earthquake Engineering and Structural Dynamics* 31, no. 3, pp. 491–514.

Wilson, E.L. 2002. *Three-Dimensional Static and Dynamic Analysis of Structures*, 3rd ed. Computer and Structures. Berkeley, California.

Zovani, E.H., and R.M. Barrionuevo. 2004. "Response to Orthogonal Components of Ground Motion and Assessment of Percentage Combination Rules." *Earthquake Engineering and Structural Dynamics* 33, no. 2, pp. 271–284.

8 Performance-based Design with the Endurance Time Method

8.1 INTRODUCTION

Seismic performance of structures subjected to strong earthquakes is one of the greatest concerns regarding the safety and economic requirements to be set as their design criteria.[1] Owners and investors need to know about the seismic performance of their structures in order to make relevant economic decisions with a reasonable level of confidence. This interest has led engineers to develop methods for designing structures such that they are capable of delivering a predictable performance during an earthquake. Performance-based earthquake engineering essentially consists of various procedures whereby the structure is expected to provide an acceptable seismic performance. The procedure involves the following: Identification of the hazard level for the site; development of conceptual, preliminary, and final structural designs; construction; and maintenance of the building during its lifetime, considering acceptable seismic performance (Krawinkler and Miranda 2004).

As per FEMA-302 NEHRP Recommended Provisions for the Seismic Regulations for New Buildings and Other Structures (FEMA-302 1997) and FEMA 273 NEHRP Guidelines for the Seismic Rehabilitation of Buildings (FEMA 1997), three performance levels (PLs) are conventionally considered. These are termed Immediate Occupancy (IO), Life Safety (LS), and Collapse Prevention (CP). In the first damage state, IO, only minor structural damages are visible, and no substantial reduction in building gravity or lateral resistance should occur. In the LS level, although significant damage to the structure may occur, structural elements should have enough capability of preventing collapse. The CP level is defined as the post-earthquake damage state, in which critical damage occurred, and the structure is on the verge of collapse (FEMA 1997). Another important concept in performance-based design is the notion of the "Performance Objective" (PO), which consists of the specification of a desired structural PL (e.g., CP, LS, or IO) at a given level of seismic hazard. For example, in accordance with SAC 2000 (FEMA-350), ordinary buildings are expected to provide, over 50 years, less than a 2 percent chance of damage exceeding CP performance (Krawinkler and Miranda 2004).

Evaluating the performance of existing structures undergoing an earthquake is another important task, through which the operational situation of a structure during and after the event can be predicted. The performance evaluation consists of structural analysis with computed demands on structural elements compared with specific acceptance criteria provided for each of the various PLs (Humberger 1997). These acceptance criteria are really some limitations that are specified for various structural parameters (such as interstory drift and plastic rotation at joints) at different PLs. The performance evaluation might sound like a straightforward process but, in reality, it is not a simple undertaking. The erratic nature of earthquakes, uncertainties in the existent analysis methods, and lack of enough information about the current strength of the structures are some factors that make the procedure intricate.

In this chapter, some new methodologies for extending the application of the Endurance Time (ET) method into the area of performance-based design will be provided. In the ET method, the structural responses at different excitation levels are obtained in a single time-history analysis, thus, significantly reducing the computational demand (Estekanchi, Vafai, and Sadeghazar 2004). Thus, by applying the ET method and considering the concepts of performance-based design, the performance of a structure at multiple levels of seismic hazard can be obtained in a single time-history analysis. This can be very beneficial regarding computational efficiency. In other words, one can check if the structure satisfies its POs by one time-history analysis using ET intensifying excitations. This characteristic of the ET method can best be utilized by extending the concepts of the ET method to merge with basic concepts from performance-based design, which is the main purpose of this chapter.

As will be explained in detail, in this proposed methodology, two new concepts of continuous Performance Curves (PCs) and generalized Damage Levels (DLs) are introduced. Utilizing some equations such as Gutenberg–Richter equations, the equivalent endurance times corresponding to each PL are obtained. A Target Performance Curve or Target Curve representing the required performance as a continuous function of the excitation level is drawn. The definition of the Target Performance Curve assumes that the PLs need not be defined as a limited number of discretized risk levels. Rather, at least from theoretical viewpoint, it is possible to consider an unlimited number of PLs in the form of a continuous curve. Then, in order to have a more versatile numerical presentation of the PLs, an index is introduced, called "Damage Level" (or "DL"), which is defined as a numerical index in the form of a real number. Integer numbers in this interval are representatives of codified PLs, thus, a convenient numerical equivalent is created for each PL. As will be explained, this index is used in order to draw a continuous target curve as the indicator of PLs. The actual performance is then plotted against this target performance based on the results of ET analysis. The overall performance of the structure can be anticipated by comparing the target to the actual performance, and the design can be improved based on the observed performance.

In order to show how this process works, the performance of three steel moment frames is evaluated using the target curve, and the advantages and limitations of this procedure are explained. As will be explained, by providing a good estimate of structural performance at different excitation levels in each response-history analysis, the ET method can considerably reduce the huge computational effort required for the practical performance-based design of structures. Also, the concepts of PC and DL provide a simplified presentation of performance analysis results that can be used as a tool in practical design cases.

8.2   CONCEPTS FROM THE ENDURANCE TIME METHOD

ET analysis is a dynamic procedure that predicts the seismic response of structures by subjecting them to a gradually intensifying dynamic excitation and monitoring their performance at different excitation levels. Structures that can endure the imposed intensifying acceleration function for longer are considered to be capable of sustaining stronger seismic excitations. In fact, since intensifying acceleration functions are used, the analysis time in ET analysis can be considered as a measure of the intensity (Riahi and Estekanchi 2010).

The concept of the ET method is usually presented by a hypothetical shaking table experiment. Assuming that different frames with unknown seismic properties are fixed on the table, and the table is subjected to an intensifying shaking, and considering the times at which the models have failed, one can rank these frames according to their seismic resistance. Hence, the endurance time of each structure against intensifying shaking can be considered as the seismic-resistance criterion (Estekanchi, Valamanesh, and Vafai 2007).

In Figure 8.1, the schematic result of a sample hypothetical experimental ET analysis is presented. The demand and capacity ratio has been calculated for these frames as the maximum absolute value of the endurance index during the time interval from 0.0 to t. If a structure is assumed to be at collapse level when its demand per capacity ratio exceeds unity, the endurance time for each frame can be derived from this figure.

To start an ET time-history analysis, after a representative model has been constructed one should set a suitable damage measure and an appropriate ET accelerogram (Figure 8.2). The analysis results are usually presented by ET response curves, in which the maximum absolute value of the damage measures in the time interval $(0, t)$ – as given in Equation (8.1) – are drawn against analysis time.

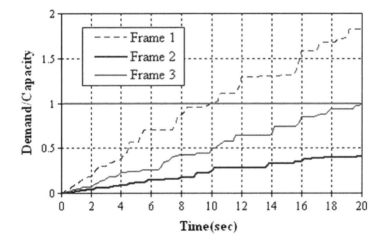

FIGURE 8.1   Demand /capacity of frames under acceleration function action (Riahi and Estekanchi 2010).

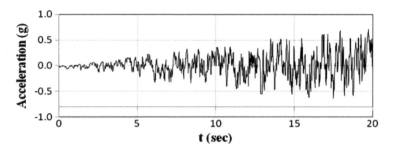

FIGURE 8.2   Typical ET accelerogram.

$$\Omega\big(f(t)\big) \equiv Max\big(Abs\big(f(\tau): \tau \in [0,t]\big)\big) \qquad (8.1)$$

Where $\Omega()$ is a Max-Abs operator and $f(t)$ is any structural response (or a function of response) such as interstory drift ratio, base shear, or damage index. For application of the ET method, intensifying accelerograms are usually generated in such a way as to produce dynamic responses equal to the scaled desired response spectrum (such as the code's design spectrum) at a predefined times (Estekanchi, Valamanesh, and Vafai 2007). If such accelerograms are used, it would be possible to compare the results of the ET time history analysis with those obtained from other analysis methods and, moreover, to compare the performance of different structures with different periods of free vibration. The first generations of suggested intensifying accelerograms for ET analysis have a linear intensification scheme – that is, the response spectrum of an ET accelerogram should intensify proportionally with time. Hence, the target acceleration response of an ET accelerogram can be related to the codified design acceleration spectrum as

$$S_{aT}(T,t) \equiv S_{aC}(T)\frac{t}{t_{Target}} \qquad (8.2)$$

where $S_{aT}(T,t)$ is the target acceleration response at time t, T is the period of free vibration, and $S_{aC}(T)$ is the codified design acceleration spectrum. Using unconstrained optimization in the time domain, the problem can be formulated as follows:

$$Find\ a_g(t)\ s.t. \forall T \in [0,\infty), t \in [0,\infty) \rightarrow \Omega\big(\ddot{u}(t)\big) = S_{aT}(T,t) \qquad (8.3)$$

Which can be formulated as a numerical optimization problem as in formula (8.4):

$$Minimize\ F\big(a_g\big) = \int_0^{T_{max}}\int_0^{t_{max}}\big\{Abs\big[S_a(T,t) - S_{aT}(T,t)\big] + \alpha$$
$$\times Abs\big[S_u(T,t) - S_{uT}(T,t)\big]dtdT\big\} \qquad (8.4)$$

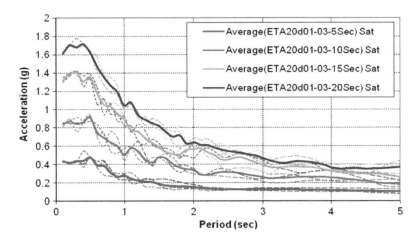

FIGURE 8.3   Acceleration response generated by ET accelerograms at various times.

Where a_g is the ET accelerogram being sought, $S_{aT}(T,t)$ and $S_{dT}(T,t)$ are the target acceleration response and displacement response at time t, respectively, $S_a(T,t)$ and $S_d(T,t)$ are the acceleration response and displacement response of acceleration function at time t, respectively, α is a weight factor, and T is the period of free vibration (Riahi and Estekanchi 2010).

It should be noted that based on the mentioned linear scheme, different sets of ET accelerograms can be generated according to their compatibility with different seismic spectra and in different response ranges (linear or nonlinear). Each set consists of a group of intensifying acceleration functions (usually consisting of three excitation functions). For example, three acceleration functions, named "ETA20d01-03" or, briefly, "d series" are in this study created in such a way that its response spectrum at $t = 10$ sec would be compatible with INBC 2800 design spectrum.

Sample response spectra generated using various time windows of ETA20d accelerograms are shown in Figure 8.3. In this figure, the curves are taken as average values between the results of three accelerograms of the d series. As can be seen in this figure, the response spectra produced by the ET acceleration function proportionally grows with time.

8.3   PERFORMANCE LEVELS

Performance Levels (PLs) are structural damage states that must be clearly defined as one of the first steps in a performance-based design procedure. These levels are usually expressed as some distinct bands in the damage spectrum of a structure, and they divide damage status of structures according to the amount of damage to structural and nonstructural components. Moreover, some other concepts, such as cost, repair time, and injury, can also be related to PLs (Grecea, Dinu, and Dubina 2004). Noting that the PLs are usually investigated at some specific levels of design earthquake motion, they can be thought of as a criterion for limiting values of measurable

TABLE 8.1
Selected Performance Objective for a Residential Building

Earthquake Having Probability of Exceedance	Mean Return Periods (Years)	Performance Level
50% per 50 year	75	IO
10% per 50 year	475	LS
2% per 50 year	2,475	CP

structural response parameters (such as interstory drift and absolute acceleration and displacement), under each considered level of earthquake motion.

There are three well-known PLs considered by FEMA-273 (NEHRP Commentary on the Guidelines for the Seismic Rehabilitation of Buildings 1997): IO, LS, and CP, which were determined according to structural damages observed in earthquakes. For example, at the IO level, the building has experienced limited damage, and at the CP level, damage is relatively extreme. On the other hand, in FEMA-356 Prestandard and Commentary for the Seismic Rehabilitation of Buildings, four levels of probabilistic earthquake hazard are defined. Combining these levels with the PLs, a table of POs can be created (FEMA-356 2000). These objectives are different according to the type of structure to be considered. For example, if a hospital is planned to be built, an appropriate PO might be that it is capable of meeting the LS PL in an earthquake with a mean return period of 2,475 years, and the IO PL in an earthquake with a mean return period of 475 years. So, if one wants to evaluate the performance of a specific model of a hospital, he or she should first analyze the model under two sets of considered earthquakes with the defined mean return periods separately, and then see if the model satisfies the related code criteria for each PL.

Since the descriptions of the POs are mostly qualitative, some performance criteria have been defined to bind these descriptions to engineering demand parameters so the POs can be predicted in analysis and design process (Krawinkler and Miranda 2004). In fact, these criteria are the rules and guidelines that must be met to ensure that the designed structure satisfies the POs. In this chapter, for explanatory purposes, the POs shown in Table 8.1 are considered to be for a residential building.

8.4 TARGET AND PERFORMANCE CURVES

As previously mentioned, the target curve is a concept by means of which the specific properties of the ET method can be readily put into use in the performance-based design procedure. Using the target curve is an appropriate way to evaluate the performance of structures in ET method. In fact, the target curve will be used as the criteria curve for the ET response curve, and the performance of a structure at different excitation levels can be evaluated by comparing the ET response curve with the target performance curve.

Since, in the target performance curve (or simply target curve), the target and the ET response curves will be compared in a single chart, the horizontal and vertical axes

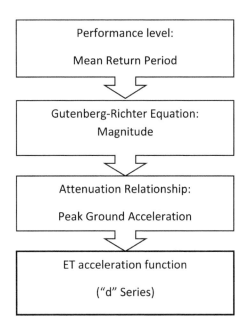

FIGURE 8.4   Target curve construction procedure.

of the target curve should be defined so as to match with the corresponding axes in the ET response curve; that is, the analysis time on the horizontal axis and a damage index on the vertical axis. The performance criteria are conceptually similar to the damage indices. Thus, the challenge is to correlate the PLs to the ET analysis time. In other words, the first step in the process of creating the target curve is to identify the respective endurance time at each PL. To do so, the procedure shown in Figure 8.4 can be followed in a step-by-step manner. It should be noted that this procedure is not a very precise one. To have a more precise calculation, it is necessary to obtain the hazard curve for a special site, and the PGA should be acquired according to that curve, but, since the main purpose in this chapter is to illustrate the basic concepts of the proposed procedure, an approximate procedure can be considered as appropriate. The corresponding magnitude of each earthquake hazard has been obtained first using the Gutenberg-Richter relation, as follows (Mohraz and Sadek 2003):

$$Log(N) = a - bM \tag{8.5}$$

where

N is the return period of earthquake (years);
M is the magnitude of earthquake in Richter;
b is called the "b-value," and is typically in the range of 0.8–1.2;
and a is called the "productivity."

Parameters "*a*" and "*b*" depend on the properties of the region in which the earthquake occurred. For Iran, the following form is recommended by Kaila and Narian (1971):

$$Log(N) = 6.02 - 1.18M \qquad (8.6)$$

After that, the peak ground acceleration (*PGA*) must be acquired from magnitude. The following formula, suggested for Iran, will be used as an example (Amiri, Mahdavian, and Manouchehri Dana 2007):

$$Ln(PGA) = 3.65 + 0.678M - 0.95Ln(R) \qquad (8.7)$$

where "*PGA*" is peak ground acceleration in cm/sec^2, "*M*" is magnitude in Richter and "*R*" is the distance to the fault in km, which is to be considered as 18 for the example in this chapter.

This acceleration can be easily related to the ET analysis time. For this purpose, it is needed to specify which series of accelerograms are to be used. According to its conformability with the Iranian code 2800 standard (BHRC 2005), the "d" series of accelerograms (i.e., ETA20d01-03) has been chosen here. Considering this series, the equivalent analysis times in ET records corresponding to the three mentioned PGAs can be identified. To do so, one method is to trace the times in the ET acceleration function at which the PGAs exceed the values of the PGAs corresponding to each PL.

The results of the previous procedure are shown in Table 8.2. In this table, the relevant interstory drift for each PL is indicated. These quantities are for steel moment frames and based on the FEMA-356 standard. Although these values are not intended in FEMA-356 to be used as acceptance criteria for evaluating the performance of structure –, and they are just some quantities that qualitatively indicate the behavior of structures at each level of performance – in this research, these values will be used as an index to show the limits of each PL. Plastic rotation in beams is the other index, which is used here to evaluate the performance of structures. In order to more accurately evaluate the performance, one should obtain the values of plastic deformations in all elements (including beams, columns, panel zones, braces, etc.) and compare them to the acceptance criteria given in the FEMA-356 standard.

TABLE 8.2
Endurance times related to each PL

Performance Level	Mean Return Periods (years)	Magnitude (Richter)	PGA(g)	Endurance Time (Sec)	Interstory Drift (%)
IO	75	6.7	6.7	5.16	0.7
LS	475	7.3	7.3	10.16	3.5
CP	2,475	7.9	7.9	15.46	5

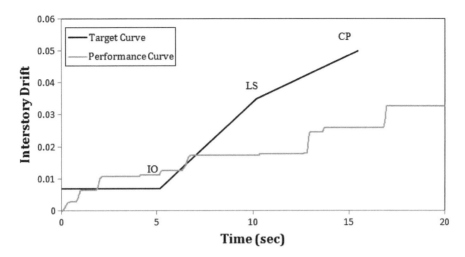

FIGURE 8.5   Target and existing performance curves.

This means that the design of a typical structure should be such that if the ETA20d series of accelerograms were applied to the structure, it would be capable of meeting the IO PL up to 4.14 sec, the LS PL up to 10.21 sec, and the CP PL up to 19.11 sec.

Based on earlier discussion, the target curve has been drawn and compared with a three-story frame performance curve in Figure 8.5. In this figure, the limit of each Performance Level (PL) has been specified on the target curve. This frame is subjected to the ETA20d series of ET accelerograms, and its interstory drift response is considered as the damage index. It should be noted that, while there are no criteria for damages below the IO level, this area is restricted by a horizontal line in the target curve.

8.5   DEFINITION AND APPLICATION OF DAMAGE LEVELS

The previous target curve has some limitations and can be improved for presentation purposes. One problem is that, in order to evaluate the performance of structures, the performance of all elements should be considered by observing their plastic deformations and comparing the values with the acceptance criteria. Since different limits are set on these parameters for various elements in each PL, it is difficult to compare the performance of different elements and readily identify the critical one. Thus, in order to evaluate the performance of a structure, one should create a target curve for each mentioned index and element and compare the related response curve with that target curve. In this way, even though the performance can be identified, the specification of the critical index is not a simple matter. A combined damage indicator can be defined for this purpose that simplifies the compilation of DLs indicated by various indexes into one normalized numerical value. Another property of the DL is that this dimensionless index creates a numerical presentation for PLs – that is, one could distinguish and compare the performance of different structures with only one number. Moreover, if two structures lie within the same PL, their performance can be still comparable with this index.

To specify such an index, five PLs are considered as OP (fully operational), IO, LS, CP, and CC (complete collapse, an arbitrary point to extend the target performance curve beyond CP), which is a rather arbitrary level to simplify formulation. It should be noticed that OP and CC levels are used not just as the limits of the performance but also the limits of the *DL*. Until more research is available to define the CC point based on more rational criteria, it will be considered arbitrary in such a way that the slope of the target performance curve before the CP level, is maintained. This additional point is required in order to theoretically extend the performance curve beyond CP and has no practical significance in this study. The formulation proposed for the *DL* has been arranged is such a way as to assign an explicit number (preferably an integer) to each PL and use the determining parameters (such as interstory drift and plastic rotation) to compute the *DL* in a clear and understandable way.

An appropriate formulation, which satisfies the earlier-mentioned considerations, can be expressed as follows:

$$DL = \sum_{i=1}^{n} \frac{\max[\theta_{i-1}, \min(\theta, \theta_i)] - \theta_{i-1}}{\theta_i - \theta_{i-1}}$$ (8.8)

where θ is the considered parameter (such as drift or plastic rotation), which should be computed from analysis, and θ_i is the considered limit (e.g. per FEMA-356 in this study) of that parameter for each PL. The values of θ_i for interstory drift and plastic rotation for each PL and the corresponding DLs are given in Table 8.3. As can be seen in Table 8.3, the obtained *DL* will satisfy its purpose satisfactorily. Because, first, it denotes the PL of the structure; second, since it is a number, it will satisfy the need to

---

TABLE 8.3
Assigning Damage Levels to Performance Levels

Performance Level	Damage Level	θ_i (Drift)	θ_i^* (Plastic Rotation)		
			Case (c)	Case (b)	Case (a)
IO	1	0.7	1	0.25	Interpolation
LS	2	3.5	6	2	required
CP	3	5	8	3	
CC**	4	7	11	6	

Notes:
* Case (a): $b_f/2t_f < 52/\sqrt{F_{ye}}$ and $h/t_w < 418/\sqrt{F_{ye}}$,
  Case (b): $b_f/2t_f > 65/\sqrt{F_{ye}}$ or $h/t_w > 640/\sqrt{F_{ye}}$,
  Case (c): Other.

** CC is an auxiliary point included so that the performance curve can be theoretically extended beyond CP

create a numerical presentation for PLs; and, third, it can present all parameters in a normalized form that will ease future comparison and computations.

In light of earlier discussions, the determining parameter, like interstory drift (or plastic rotations) can be replaced by DL in the target curve. Likewise, the structure performance or response curve can be drawn according to this new index to comply with the target curve.

8.6 STRUCTURAL MODEL FRAMES DESIGN

In order to demonstrate how the target curve can ease the visualization of the performance of structures, a set of 2D steel moment-resisting frames with a different number of stories and spans were selected and used in this chapter. These models consist of 3-story, 1-bay and 7-story, 3-bay frames that are designed in 3 alternatives (standard, weak, and strong frames) and a 12-story, 3-bay frame that is designed in 2 standard and strong alternatives. These frames are designed according to the AISC-ASD89 design code. To compare the performance of the frames with varying seismic resistances, the standard, weak, and strong frames have been designed using base shears equal to 1, 0.5, and 1.5 times the codified base shear, respectively. As an example, the geometry and section properties of the seven-story, three bay frames are shown in Figure 8.6. In this figure, the black circles stand for the plastic hinges and show that the failure mechanism follows the strong column–weak beam concept. As can be seen in Figure 8.7, these hinges have been modeled as a rotational spring, with a moment-rotation curve shown in this figure. The capital letters in this figure (A to E) determine the boundaries of various behaviors of the hinge model.

8.7 PRESENTATION OF THE ANALYSIS RESULTS

The modeling and nonlinear analysis were done with OpenSees software (PEERC 2004) [14]. Nonlinear models of the frames are prepared by using the beam element with nonlinear distributed plasticity. In this study, the damping ratio is assumed to be 0.05 of the critical value and P-Δ effects have been included in the nonlinear analysis. Applicability of the ET method in nonlinear analysis and acceptability of its approximation in estimating various damage indexes have already been studied (Riahi and Estekanchi 2010; Riahi, Estekanchi, and Vafai 2009; Estekanchi, Arjomandi, and Vafai 2008). A similar level of approximation is considered to be adequate for the purpose of this study.

The drift and plastic rotation responses of frames were obtained and converted to the DL index, according to Equation 8.8. Then, the ET response curve (performance curve) of each frame was plotted separately for each aforementioned parameter's related DL. In Figure 8.8, the response curve for the plastic rotation and drift in a three-story standard frame is depicted. As illustrated in this figure, using the concepts of ET and DL, it is possible to compare the situation of various parameters, such as plastic rotation and drift, and identify the critical parameter in any seismic intensity. For example, in a three-story standard frame, as can be seen in Figure 8.8, drift is the critical parameter in low-intensity ground motions, but plastic rotation is critical

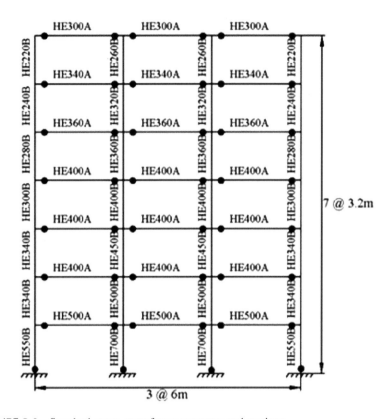

FIGURE 8.6 Standard seven-story frame geometry and sections.

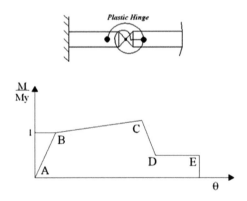

FIGURE 8.7 Plastic hinge model and its generalized force-deformation curve.

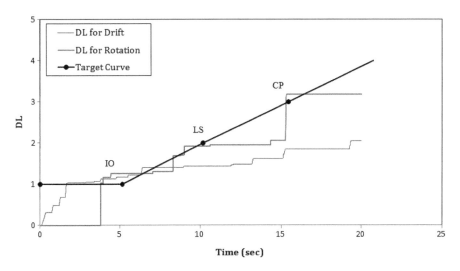

FIGURE 8.8 Performance curves for plastic rotation and drift in frame F3s1b.

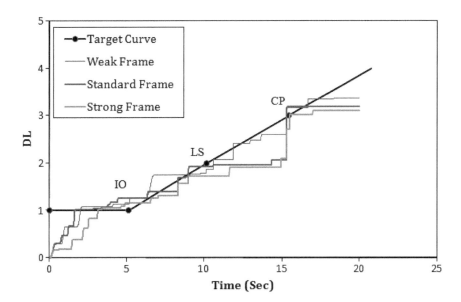

FIGURE 8.9 Performance curves for three-story frames.

at high-intensities. Note that the final performance curve for each frame should be created considering the maximum value of DL considering both the drift and rotation (or any other parameters that need to be considered based on the design criteria). The final DL response curve (or performance curve) for three alternatives of a three-story frame is shown in Figure 8.9. Using this figure, one can easily study and compare the performance of these three alternatives in various seismic intensities. For example,

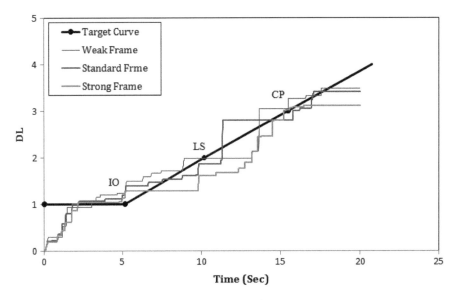

FIGURE 8.10   Performance curves for seven-story frames.

according to Figure 8.9, it can be concluded that all three alternatives fail the criteria of the IO PL, but remain in the safe region of LS and CP PLs; the standard frame performs similarly to the weak frame at low intensities, but its performance resembles the strong-frame performance as seismic intensity increases.

Figure 8.10 shows the performance curves of three types of seven-story frames. A good performance of the strong frame in comparison with two other frames can be easily observed in this figure. Moreover, it can be seen that the weak frame lies above the target curve almost at all times. The behavior of 12-story frames can be evaluated with a similar procedure.

Using the target curve, the performances of strong frames with various numbers of stories and bays can also be compared with each other. Such a comparison has been done for three strong frames, with 3, 7, and 12 stories (Figure 8.11). Although all three frames were designed on the basis of 1.5 times the codified base shear, their performance is not similar. In this study, the 12-story frame is the best performer, and the 3-story frame is the worst.

To show and verify the versatility of the target curve, the three-story standard frame has been subjected to some earthquake records, and its performance is evaluated by the target curve. To have a good evaluation, seven earthquake records have been selected from the FEMA-440 recommended records and scaled in such a way as to create seven modified records at each PL (i.e., 21 records). To do so, some correction factors (i.e. scale factors) are required to be computed and applied to each record. The correction factors for each PL and each record are the ratio of PL-related PGA (see Table 8.2) to the record's PGA. In Table 8.4, the properties of used records and the mentioned correction factors for each PL are provided. In this way, 21-time history analyses have been performed, and responses of the frame are calculated in the form

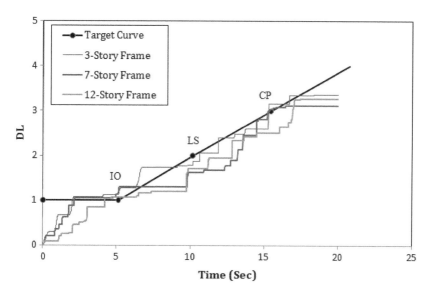

FIGURE 8.11   Performance curves for strong frames.

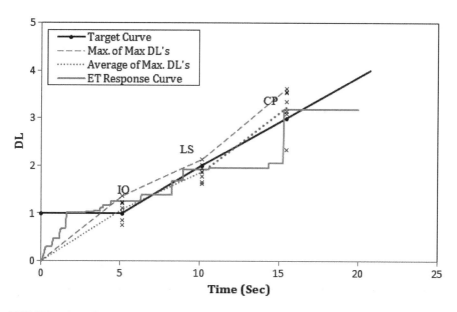

FIGURE 8.12   Comparison of ET and earthquakes results at equal PGA for three-story standard frame.

of the DL index. To provide a logical overview of the performance of the standard frame, it is recommended that the maximum and average of the DL indices in each PL be considered and compared to those related to ET analysis results. Figure 8.12 shows the aforementioned comparison. In this figure, the crosses stand for the results

TABLE 8.4

Properties of Used Records and the Correction Factors

No	Record	HP	LP	DT	Name	CF for CP	CF for LS	CF for IO	PGA
1	Northridge 24278	0.12	23	0.02	NRORR360	4.035	2.222	1.053	0.171g
2	Landers 12149	0.07	23	0.02	LADSP000	2.664	1.467	0.695	0.259g
3	Loma Preita 1652	0.2	41	0.005	LPAND270	1.575	0.868	0.411	0.438g
4	Loma Preita 47006	0.2	45	0.005	LPGIL067	2.255	1.242	0.588	0.306g
5	Morgan Hill 57383	0.1	27	0.005	MHG06090	1.933	0.504	0.504	0.357g
6	Loma Preita 58065	0.1	38	0.005	LPSTG000	2.828	1.577	0.615	0.244g
7	Loma Preita 58135	0.2	40	0.005	LPLOB000	1.342	0.739	0.35	0.514g

of 21-time history analyses. As can be seen in this figure, the average values are compatible with ET analysis results – that is, the performance estimated by the ET method is consistent with those from time history analysis using ground motions. Now it seems that if a line connects the average (or maximum) values, the performance of the frame will be coincident with this line at various seismic intensities. In fact, this line is similar to the familiar performance curve. However, it should be mentioned that selecting the optimal form of the connecting curve requires further research in this area.

8.8 SUMMARY AND CONCLUSIONS

In this chapter, some methodologies are proposed in order to extend the application of the ET method to performance-based design of structures. Application of the ET method in performance-based design of steel frames is investigated from a conceptual viewpoint. In the ET method, structures are subjected to an intensifying acceleration function, thus, an estimate of the structural response at different levels of excitation is obtained in a single response history analysis, thus, considerably reducing the required computational effort in multilevel performance assessment. The concept of PLs has been extended from discrete presentation to a continuous target performance curve. This target performance curve, while theoretically more attractive, turns out to be quite versatile when investigating the ET analysis results. In order to be able to combine several different performance criteria into a single numerical performance index, a generalized DL index has been proposed. The DL index proposed in this chapter creates a versatile numerical representation of PLs and also provides a uniform index to express a performance of structures that incorporates various parameters (such as drift, plastic rotation, etc.). Furthermore, the target curve is an effective tool for estimating the performance of structures under various seismic intensities by the ET response curve. This curve can be used to anticipate the seismic performance of structures subjected to earthquakes. The target performance curve has good potential to be used in the performance-based design of structures. The concepts

of Performance Curve and DL introduced in this chapter lay the necessary founda-
tion for a more versatile application of the ET method in the practical performance-
based design of structures. The analysis results are shown to be consistent with those
obtained using ground motions scaled to represent particular excitation levels. It
should be noted that more research in this area is required in order to assess the precision and reliability that can be expected from proposed methodology.

8.9 NOMENCLATURE

Abs	= Absolute value function
$a_g(t)$	= ET acceleration function
b_f	= Flange width
CC	= Complete collapse
CP	= Collapse prevention performance level
DL	= Damage level
$F(a_g)$	= Optimization target function
F_{ye}	= Yield strength
IO	= Immediate occupancy performance level
LS	= Life safety performance level
M	= Earthquake magnitude
Max	= Maximum of the values
N	= Earthquake mean return period
PGA	= Peak ground acceleration
PL	= Performance level
R	= Distance to the fault
S_a	= Spectral acceleration
$S_a(T,t)$	= Acceleration response for period T at time t
$S_a(T)$	= Acceleration response as a function of period T
$S_{aC}(T)$	= Codified design acceleration spectrum
$S_{aT}(T,t)$	= Target acceleration response for period T at time t
$S_d(T,t)$	= Displacement response value for period T at time t
$S_{dT}(T,t)$	= Target displacement response value for period T at time t
T	= Free vibration period (sec)
t	= Time
t_f	= Flange thickness
t_{Target}	= Target time (=10 sec in this study)
T_{max}	= Maximum free vibration period (sec) to be considered in the optimization
t_{max}	= Time corresponding to the end of the acceleration function
α	= Weighting factor in optimization target function (=1.0 in this study)
θ_i	= Determinant parameter (drift ratio or plastic rotation)
$\Omega(f(t))$	= ET analysis result equal to $\text{Max}(\text{Abs}(f(t1)))$: $t1 \in [0,t]$
|	= Such that
$\forall$	= For all values

NOTE

1 *Chapter Source*: Mirzai, A., H.E. Estekanchi, and A. Vafai. 2010. "Application of Endurance Time Method in Performance Based Design of Steel Moment Frames," *Scientia Iranica* 17, no. 6, pp. 482–492.

REFERENCES

Amiri, G.G., A. Mahdavian, and F. Manouchehri Dana. 2007. "Attenuation Relationships for Iran." *Journal of Earthquake Engineering* 11, no. 4, pp. 469–492.

BHRC 2005 *Iranian Code of Practice for Seismic Resistance Design of Buildings: Standard No. 2800.* 3rd Edition, Building and Housing Research Center. Tehran (In Persian)

Estekanchi, H.E., K. Arjomandi, and A. Vafai. 2008. "Estimating Structural Damage of Steel Moment Frames by Endurance Time Method." *Journal of Constructional Steel Research* 64, no. 2, pp. 145–155.

Estekanchi, H.E., A. Vafai, and M. Sadeghazar. 2004. "Endurance Time Method for Seismic Analysis and Design of Structures." *Scientia Iranica* 11, no. 4, pp. 361–370.

Estekanchi, H.E., V. Valamanesh, and A. Vafai. 2007. "Application of Endurance Time Method in Linear Seismic Analysis." *Engineering Structures* 29, no. 10, pp. 2551–2562.

FEMA. 1997. *NEHRP Guidelines for the Seismic Rehabilitation of Buildings.* FEMA-273, Washington, DC: Federal Emergency Management Agency.

FEMA-302. 1997. *NEHRP Recommended Provisions for Seismic Regulations for New Buildings and Other Structures, Part 1: Provisions.* Washington, DC: Prepared by the Building Seismic Safety Council for the Federal Emergency Management Agency.

FEMA-350. 2000. *Recommended seismic design criteria for new steel moment-frame buildings.* Rep. No. FEMA-350, SAC Joint Venture, Washington, D.C.

FEMA-356. 2000. *Prestandard and Commentary for the Seismic Rehabilitation of Buildings.* Washington, DC: Federal Emergency Management Agency.

Grecea, D., F. Dinu, and D. Dubina. 2004. "Performance Criteria for MR Steel Frames in Seismic Zones." *Journal of Constructional Steel Research* 60, no. 3, pp. 739–749.

Humberger, R.O. 1997. "A Framework for Performance-Based Earthquake Resistive Design." *EERC-CUREe Symposium in Honor of Vitelmo V. Bertero*, Berkeley, CA.

Kaila, K.L., and H. Narian. 1971. "A New Approach for the Preparation of Quantitative Seismicity Maps." *Bulletin of the Seismological Society of America* 61, no. 5, pp. 1275–1291.

Krawinkler, H., and E. Miranda. 2004. "Performance-Based Earthquake Engineering." In *Earthquake Engineering: from Engineering Seismology to Performance-Based Engineering*, Y. Bozorgnia and V.V. Bertero, eds, Chapter 9, pp. 1–59. Boca Raton: CRC Press.

Mohraz, B., and F. Sadek. 2003. "Earthquake Ground Motion and Response Spectra." In *The Seismic Design Handbook*, F. Naeim, ed. Springer, Boston, MA.

Pacific Earthquake Engineering Research Center (PEERC). 2004. *Open System for Earthquake Engineering Simulation.* (OpenSees), Berkeley, CA: Pacific Earthquake Engineering Research Center (http://opensees.berkeley.edu/).

Riahi, H.T., and H.E. Estekanchi. 2010. "Seismic Assessment of Steel Frames with Endurance Time Method." *Journal of Constructional Steel Research* 66, no. 6, pp. 780–792.

Riahi, H.T., H.E. Estekanchi, and A. Vafai. 2009. "Application of Endurance Time Method in Nonlinear Seismic Analysis of SDOF Systems." *Journal of Applied Sciences* 9, no. 10, pp. 1817–1832.

9 Value-based Seismic Design with ET

9.1 INTRODUCTION

In recent decades, large economic losses following earthquakes and hurricanes have shown the need for improved design criteria and procedures that provide the necessities to reduce damage and economic impacts to an acceptable level along with protecting lives.[1] The prescriptive and performance-based approaches of seismic design try to find a structure to satisfy minimal requirements under seismic actions in a number of levels of intensity, and a design with lower initial cost is normally preferred. These approaches will not necessarily result in an economical design with lower total cost over the lifetime of the structure. The goal of achieving the most economical design over the entire lifetime of the structure can be considered as a design value. While other goals, such as resilience, environmental impact, sustainability and other criteria, can be defined in the framework of Value-based Seismic Design, the design's overall economy can usually be considered the most relevant and popular measure of value. Also, most design goals can be interpreted in terms of their equivalent economic value. While, in this chapter, value is assumed to consist of maximum economic value, it should be clear that the concept of the value in the framework of Value Based Seismic Design can be extended to cover a broader range of design goals as well (Mirfarhadi and Estekanchi 2020). In order to incorporate the economic concerns in design or decision-making procedures more directly, Life-cycle Cost Analysis (LCCA) has been applied in the construction industry. LCCA has provided a reliable tool for estimating costs due to future earthquakes during the design life of a structure. This analysis, in companion with an optimization algorithm, can result in a design with the least total cost. The LCCA principles are based on economic theories and was mainly implemented to introduce financial concerns in the structural design area. However, this analysis can provide a baseline to incorporate technical, economic, and social or any other intended measures thought to be important in design procedure. Considering economic and technical issues in the design and construction field will lead to the optimal allocation of valuable resources. Although in the construction industry LCCA was first introduced in economic investment assessments of infrastructures, LCCA nowadays is an essential component of the design process used to control initial and future costs of buildings in seismically active regions and is widely used in risk assessment and decision analysis. Using this method, the expected total cost of a structure, including the initial cost and also losses resulting from earthquakes during its life span, can be considered as the main indicator of the priority of design alternatives. In this chapter, LCCA is used to determine the total value of a structure as an investment appraisal tool to be incorporated in design procedure. Readily introduced value-based design can provide a wider description of the design target by defining the earthquake consequences, such as structural damages, loss of contents, losses due to downtime, and human injuries and fatalities in the form of quantifiable parameters. In this way, it is expected that the resultant design will perform with desired post-earthquake capabilities with manageable disruption.

LCCA demands the calculation of the cost components related to the performance of the structure in multiple hazard levels (Mitropoulou, Lagaros, and Papadrakakis 2011). In order to have a reasonably reliable performance assessment and estimate, the seismic capacity of a structural system to be incorporated into the LCCA methodology, response-history-based incremental analyses, and consideration of a realistic numerical model of the structure are inevitable. However, these procedures require repetitive and massive analyses, and the huge computational demand and sophistications involved may make optimization algorithms impractical, or the simplifications used may decrease the reliability of the outcome. In this chapter, the Endurance Time (ET) method, as a dynamic procedure requiring reasonably reduced computational effort, is applied to estimate the performance of the structure at various hazard levels (Estekanchi, Vafai, and Sadeghazar 2004). In the ET method, structures are subjected to specially designed intensifying acceleration functions, and their performance is assessed based on their response at different excitation levels correlated to specific ground-motion intensities by each individual response-history analysis. Thus, the required huge computational demand of a complete response-history analysis is considerably reduced while maintaining the major benefits of it – that is, accuracy and insensitivity to model complexity (Estekanchi and Basim 2011). Application of the ET method in combination with the concept of LCCA can pave the way for practical Value-based Seismic Design of Structures (VBSD).

The ET method introduced by Estekanchi, Vafai, and Sadeghazar (2004) as an analysis method can be utilized to assess seismic performance of the structures in a continuous range of seismic-hazard intensities. In this method, structures are subjected to a predesigned, gradually intensifying accelerogram, and the seismic performance of the structure can be monitored while the seismic demand is increasing. Application of the ET method in performance assessment of structures has been studied by Mirzaee, Estekanchi, and Vafai (2010) and Hariri-Ardebili, Sattar, and Estekanchi (2014). Reasonably accurate estimates of expected seismic response at various excitation intensities of interest have been obtained through ET analysis by correlating the dynamic characteristics of intensifying excitations with those of ground motions at various hazard levels (Mirzaee and Estekanchi 2013).

In order to demonstrate the method, a five-story and three-bay steel special moment-resisting frame is optimally designed according to the Iranian National Building Code (INBC), which is almost identical to the ANSI/AISC360 (2010) LRFD design recommendations. Also, the frame is designed optimally to conform to FEMA-350 (2000) limitations as performance-based design (PBD) criteria. The performance of the designed frames is investigated by the ET method and, as a third step, a new design has been achieved using the proposed method to have the minimum total cost during a lifetime of 50 years. The resultant prescriptive, performance-based and value-based designs of the frame are different due to their distinct basic design philosophies. Seismic performance and cost components of these structures are investigated and discussed.

9.2   BACKGROUND

Although significant progress has been made in the last two decades in the area of seismic engineering, most of the seismic design codes currently belong to the category of prescriptive design codes, which consider a number of limit state design controls to provide safety. The two common limit state checks are serviceability and ultimate strength. It is worthy to note that designs with a lower weight, or initial material costs, are commonly preferred. Prescriptive building codes do not provide acceptable levels of a building's life-cycle performance, since they only include provisions aiming at ensuring adequate strength of structural members and overall structural strength (Mitropoulou, Lagaros, and Papadrakakis 2011). To fulfill the deficiencies of the primitive design procedures, design codes are migrating from prescriptive procedures intended to save lives, to a reliability-based design, and most of them have attempted to advance their design criteria toward a PBD of structures. The 1995 report of the SEAOC committee can be considered the start of this progress. Performance-based earthquake engineering affirms the methodology by which structural-design criteria are expressed in terms of achieving a set of different performance objectives defined for different levels of excitations where they can be related to the level of structural damage. In this methodology, the performance of the building after construction is inspected in order to ensure reliable and predictable seismic performance over its lifetime. In PBD, more accurate and time-consuming analysis procedures are employed in order to estimate the entire nonlinear structural response. Various guidelines on the PBD concept have been introduced over the last ten years for assessment and rehabilitation of existing structures and the analysis and design of new ones. FEMA-350 (2000) supplies a probability-based guideline for PBD of new steel moment-resisting frames considering uncertainties in seismic hazard and structural analyses. Design codes based on reliability of performance are useful in providing safety margins for the performance objectives with quantifiable confidence levels considering various sources of uncertainties.

In PBD, after selecting the performance objectives and developing a preliminary design, the seismic response of the design is evaluated. Afterwards, the design can be revised until the acceptance criteria for all intended performance objectives are satisfied. In order to achieve optimal structural designs with acceptable performance, optimization methods have been effectively used for PBD. The structural performances and structural weight are incorporated as objective functions or constraints to the optimization problem (Ganzerli, Pantelides, and Reaveley 2000). Among many others, Pan, Ohsaki, and Kinoshita (2007) incorporated multiple design requirements into a multi-objective programming problem using a new formulation based on the constraint approach, and Liu and colleagues (2013) utilized a PBD approach for a multi-objective optimization using a genetic algorithm subjected to uncertainties and provided a set of Pareto-optimal designs.

None of the prescriptive and PBD procedures has the capability of incorporating the economic issues in the design process. As a result, in conventional engineering practice, a design alternative with a lower initial cost is normally adopted. Large economic losses following recent earthquakes and hurricanes encouraged researchers to include financial criteria in the structural design. The LCCA principles are based on economic theories, and were mainly implemented for energy- and water-conservation projects as well as transportation projects. However, LCCA has become an important part of structural engineering to assess the structural comeback and evaluate the performance of the structure during its life span in economic terms. LCCA has gained considerable attention in decision-making quarters to decide on the most cost-effective solutions related to the construction of buildings in seismic regions. First, LCCA was applied in the commercial area and, in particular, in the design of products. Later in the early 2000s, as one of the impressive works in this area, Wen and Kang (2001a) formulated long-term benefit versus cost considerations for evaluation of the expected life-cycle cost of an engineering system under multiple hazards. Many subsequent works have been carried out to take account of economics in structural engineering. Liu, Burns, and Wen (2003) defined a multi-objective optimization problem and an automated design procedure to find optimal design alternatives. Static pushover analyses were done to verify the performance of steel frame design alternatives, and a genetic algorithm was used. Takahashi, Kiureghian, and Ang (2004) formulated the expected life-cycle cost of design alternatives using a renewal model for the occurrence of earthquakes in a seismic source, which accounts for the temporal dependence between the occurrence of characteristic earthquakes. They applied the methodology to an actual office building as a decision problem. Liu, Burns, and Wen (2005) formulated performance-based seismic design of steel frame structures as a multi-objective optimization problem, considering the seismic risk in terms of maximum interstory drift. Fragiadakis, Lagaros, and Papadrakakis (2006) used pushover analysis to compare single-objective optimal design of minimizing the initial weight, and two performance-based objective designs of a steel moment-resisting frame. In particular, a framework to generate a Pareto front for the solutions was presented. Kappos and Dimitrakopoulos (2008) used cost-benefit and life-cycle cost analyses as decision-making tools to examine the feasibility of strengthening reinforced concrete buildings. Mitropoulou, Lagaros, and Papadrakakis (2010) probed the influence of the behavior factor in the final design of reinforced concrete buildings under earthquake loading in terms of safety and economy by demonstrating initial and damage cost components for each design. Mitropoulou, Lagaros, and Papadrakakis (2011) investigated the effect of the analysis procedure, the number of seismic records imposed, the performance criterion used, and the structural type on the life-cycle cost analysis of 3D-reinforced concrete structures. Furthermore, the influence of uncertainties on the seismic response of structural systems and their impact on LCCA is examined using the Latin hypercube sampling method.

9.3   APPLICATION OF THE ENDURANCE TIME METHOD

ET excitation functions are in the form of artificial accelerograms created in such a way that each time window of them from zero to a particular time produces a response spectrum that matches a template spectrum with a scale factor that is an increasing function of time. This interesting characteristic has been achieved by resorting to numerical optimization procedures in producing ET accelerograms (Nozari and Estekanchi 2011). Various sets of ET acceleration functions have been produced with different template-response spectra and are publicly available through the website of the ET method (ET method website 2014). A typical ET accelerogram used in this work, ETA40h, is depicted in Figure 9.1. These records are optimized to fit an average response spectrum of seven records (longitudinal accelerograms) used in FEMA-440 for soil type (C) as template spectrum.

As can be seen in Figure 9.2, the response spectrum of a window from $t_{ET} = 0$ to $t_{ET} = 10$ sec of used accelerogram matches with the template spectrum. Furthermore, the produced response spectra also match the template spectrum at all other times with a scale factor, thus producing a correlation between analysis time and induced spectral intensity. Hence, each ET analysis time is representative of a particular seismic intensity, and results can be more effectively presented by considering a correlation between time in an ET analysis and the equivalent hazard-return period based on code recommendations considering the fact that hazard levels are well presented by acceleration response spectra in current codes. This can result in an appropriate baseline to calculate probabilistic damage and cost.

The application of the ET method in PBD was studied by Mirzaee and colleagues (2010) introducing the "Performance Curve" and the "Target Curve," which express, respectively, the seismic performance of a structure along various seismic intensities and their limiting values according to code recommendations. These concepts were studied in Chapter 8. As discussed in Chapter 5, substituting return period or annual probability of exceedance for time in the expression of the performance will make presentation of the results more explicit and their convenience for calculate probabilistic cost will be increased.

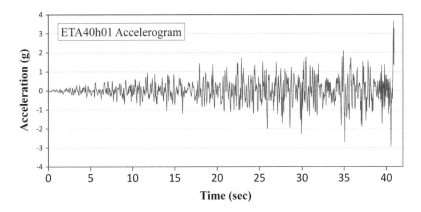

FIGURE 9.1   Acceleration function for ETA40h01.

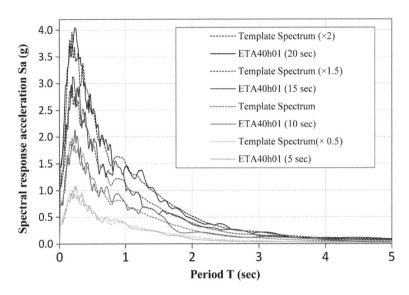

FIGURE 9.2 Acceleration response spectra for ETA40h01 at different times of excitation.

A hazard return-period corresponding to a particular time in ET analysis can be calculated by matching the response spectra at effective periods – for example, from 0.2 to 1.5 times of structure's fundamental period of vibration. The procedure is based on the coincidence of response spectra obtained from the ET accelerogram at different times and from response spectra defined for Tehran at different hazard levels. The results show that substitution of the return period for time in the ET analysis and performance curves increases the usefulness of these curves and can simplify application of the ET method in value-based design. Figures 9.3 and 9.4 illustrate the variation of the return period with the structural period and time in ET analysis. Using this correlation for a specific structure, the corresponding ET time for each hazard level is on hand. The detailed procedure to obtain such a correlation is explained in a work by Mirzaee, Estekanchi, and Vafai (2012).

In Figure 9.5 a sample target curve and performance curve for the five-story structure is depicted where ET analysis time has been mapped into the return period on a horizontal axis. As can be seen, the structure satisfies the code collapse-prevention (CP) level limitations, but it has violated the immediate occupancy (IO) and life-safety (LS) levels limitations, and the frame does not have an acceptable performance. Also, a moving average is applied to smooth ET results for the interstory drift envelope curve. It can be inferred as one of advantages of the ET method that the performance of the structure in continuous, increasing-hazard levels can be properly depicted in an easy-to-read figure.

9.4 PRESCRIPTIVE SEISMIC DESIGN

Commonly, the structure in prescriptive seismic design procedures is considered safe if it satisfies a number of checks in one or two deterministically expressed limit states

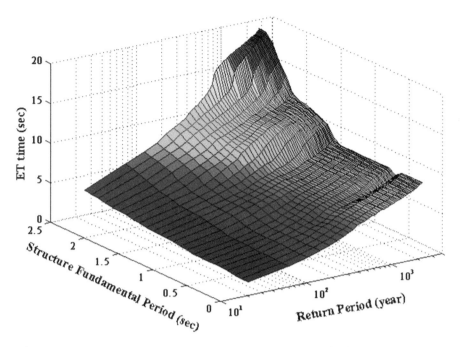

FIGURE 9.3   Return period versus structural period and ET analysis time.

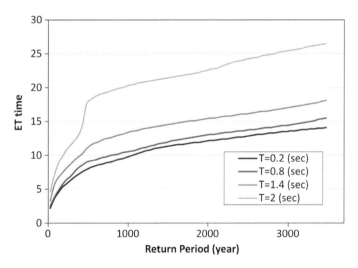

FIGURE 9.4   Equivalent ET analysis time versus hazard return period for different structural periods.

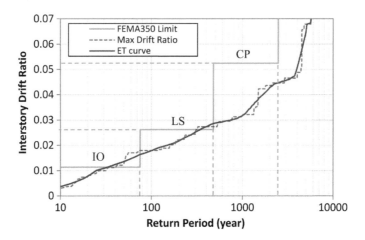

FIGURE 9.5 A sample performance curve (ET curve) for the steel frame.

(i.e., ultimate strength and serviceability). Also, the structures are allowed to absorb energy through inelastic deformation by designing them with reduced loading specified by the behavior factor leading to smaller seismic loads.

Prescriptive design of the understudy structure has been accomplished according to the Iranian National Building Code (INBC), which is almost identical to the ANSI/AISC360-10 LRFD design recommendations. The prototype structure is a five-story and one-bay special moment-resisting steel frame. All supports are fixed, and the joints are all rigid. The beams and columns are selected among seismically compact standard W profiles according to Table 9.1. The geometry of this model can be found in Figure 9.6. Loading is set according to INBC Section 6. The steel considered has a yielding stress of $F_y = 235.36$ MPa and an elastic modulus of $E = 200$ GPa. The strong column–weak beam design requirement has been considered in the design of the structure. According to Iranian seismic design code, the seismic loading base shear is determined upon design response spectrum of the 475-years return period hazard level, and the elastic base shear is reduced by a behavior factor (R) to incorporate the inelastic deformation capacity of the special moment-resisting frame. In this section, we have tried to portray the design procedure in the same way as the procedure applies in common engineering practice. Demand and capacity ratios are depicted in Figure 9.6. As can be seen in this figure, in some elements other limitations such as drift limits or the strong column–weak beam limitation, are dominant.

The seismic performance of the prescriptive design has been investigated according to FEMA-350 limitations on interstory drift ratios. The procedure and recommended limitations are explained in next section. Performance Curve and also Target Curve for this structure are depicted in Figure 9.7. It can be verified that the structure has violated IO level limitation but has a proper performance in LS and CP levels.

9.5 PERFORMANCE-BASED DESIGN

Prescriptive design procedures do not assure reliable performance of the structure in multiple-hazard levels during its life span, because these procedures merely intend

TABLE 9.1
List of Beams and Columns Alternative Section Properties

Beams	W18 ×35	W18 ×40	W18 ×46	W18 ×50	W18 ×55	W18 ×60	W18 ×65	W18 ×71						
Side columns	W10 ×39	W10 ×45	W10 ×54	W10 ×60	W10 ×68	W10 ×77	W10 ×88	W10 ×100	W10 ×112					
Inner columns	W12 ×79	W12 ×87	W12 ×96	W12 ×106	W12 ×120	W12 ×136	W12 ×152	W12 ×170	W12 ×190	W12 ×210	W12 ×230	W12 ×252	W12 ×279	W12 ×305

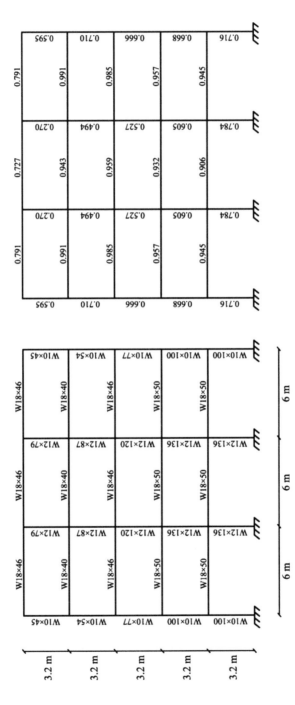

FIGURE 9.6　Schematics of steel frames under investigation and demand/capacity ratios according to prescriptive design criteria.

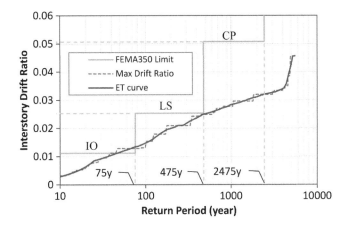

FIGURE 9.7 Performance curve (ET curve) for the prescriptive design.

to keep the ultimate strength of structural members at an acceptable level. Compared to these procedures, PBD has provided a more general structural design philosophy in which the design criteria are expressed in terms of achieving multiple performance requirements when the structure is subjected to various seismic hazard levels. In most of the current PBD criteria the structural performance of an ordinary building frame is usually defined as the following: (1) resist a significant accidental earthquake without structural damage; (2) allow repairable structural damage against a rare major earthquake; and (3) resist the maximum credible earthquake without collapse (Pan, Ohsaki, and T. Kinoshita 2007). The performance measures may include the response stresses, the maximum load carrying capacity, the interstory drift, or the plastic rotation at members, and so on. In the most common approaches to PBD, the performances against seismic motions are defined based on displacements or global deformation. Various methods of structural assessment have been used by researchers and engineers. Push-over analysis, especially, is being widely used in this area; however, time-history analysis is so far believed to be the most accurate methodology for evaluating structural performance. In order to satisfy performance-based measures in design procedure and achieve a safe yet economical design, utilizing optimization methods is inevitable. The merits of the ET method in optimum PBD and its methodology are introduced in a work by Estekanchi and Basim (2011).

PBD procedure implemented in this work is based on the method introduced in FEMA-350 (2000). This criterion supplies a probability-based guideline for PBD of new steel moment-resisting frames, in which the ground motion variability and the uncertainty in the structural analysis are explicitly considered. FEMA-350 considers two discrete structural performance levels, CP and IO, by introducing the limiting damage states for common framing elements related to these performance levels. And acceptance criteria are related to the permissible interstory drifts and earthquake-induced forces for the various elements, especially in columns. The interstory drift ratio is a commonly used measure of both structural and nonstructural damage because of its close relationship to plastic rotation demands on individual beam–column connection assemblies. As recommended in these criteria, other structural performance levels can be determined on a project-specific basis, by interpolation or extrapolation from the criteria provided for the two performance levels. For the purpose of this work, LS performance levels have been used by interpolating the IO and CP levels. LS level is a damage state in which significant damage has been sustained, although some margin remains against either partial or total collapse. The considered performance objective in this study – assuming "seismic use group I" for the prototype special moment-resisting frame structure – is IO, LS, and CP performance levels corresponding to ground-motion levels of, respectively, 50, 10, and 2 percent probability of being exceeded in 50 years.

Many uncertain factors exist that affect the behavior and response of a building, such as uncertainties in seismic hazard due to the attenuation laws employed, record-to-record variability or, on the other hand, uncertainties in structural modeling due to simplifications and assumptions used in the numerical analysis (Liu, Atamturktur, and Juang 2013). Therefore, FEMA-350 adopts a reliability-based probabilistic approach to performance evaluation that explicitly acknowledges these inherent uncertainties. These uncertainties are expressed in terms of a confidence level. A high level of confidence means that the building will very likely be capable of meeting the desired performance. Considering a minimum confidence level of 70 and 90 percent for, respectively, IO and CP performance levels, the upper bound limit for calculated interstory drift demand obtained from structural analysis would be 0.0114 and 0.0524, and interpolation will result in an upper bound of 0.0262 for the LS level.

Structural analysis has been performed by OpenSees (Mazzoni et al. 2006), where the nonlinear behavior is represented using the concentrated plasticity concept with zero-length rotational springs, and structural elements are modeled using elastic beam-column elements. The rotational behavior of the plastic regions follows a bilinear hysteretic response based on the Modified Ibarra Krawinkler Deterioration Model (Ibarra, Medina, and Krawinkler 2005; Lignos and Krawinkler 2010). Second-order effects have been considered using the P-Delta Coordinate Transformation object embedded in the platform. To capture panel zone shear deformations, panel zones are modeled using the approach of Gupta and Krawinkler (1999) as a rectangle composed of eight very stiff elastic beam-column elements with one zero-length rotational spring in the corner to represent shear distortions in the panel zone.

In this section, a single objective optimization problem is defined to find a design having the minimum initial steel material weight as the optimization objective and, according to FEMA-350 recommendations, as PBD criteria, the limitations on interstory drift demand and axial compressive load on columns and also strong column–weak beam criterion as optimization constraints. The design variables are the steel section sizes selected among standard W sections. As indicated in FEMA-350, structures should, at a minimum, be designed in accordance with the applicable provisions of the prevailing building code, such as specifications of AISC360 (2010) and AISC341 (2010). Thus, the AISC360 requirements and FEMA-350 acceptance criteria are implemented as design constraints. Optimum design sections have been determined using a GA algorithm adopted for PBD purposes using the ET method introduced in a work by Estekanchi and Basim (2011). The acquired design sections

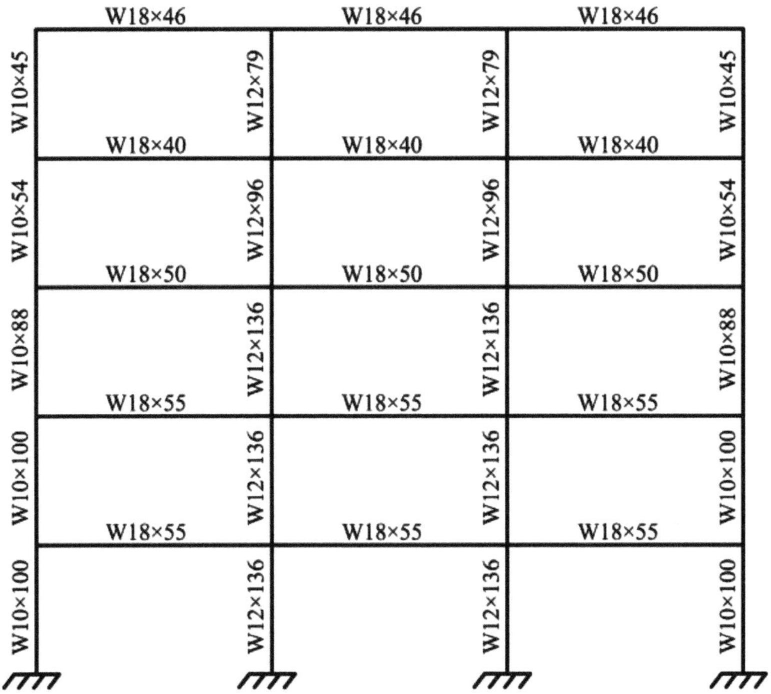

FIGURE 9.8 Performance-based design sections of the frame.

can be found in Figure 9.8. Performance of the structure in various seismic intensities can be investigated using the ET curve presented in Figure 9.9. Eventually, an optimum design will meet the constraints (i.e., code requirements) with the least margins.

9.6 VALUE-BASED DESIGN

While value can be defined and considered in its broad sense for design purposes, for the clarity of explanation, we consider in this research the structure that is more economical to construct and maintain, to be the most valued. ET analysis provides a versatile baseline to perform economic analyses on design alternatives with acceptable computational cost. Initial construction cost and expected seismic damage cost throughout the lifetime of the structure are usually the two most important parameters for decision-making (Mitropoulou, Lagaros, and Papadrakakis 2011). One of the major obstacles in seismic damage cost assessment of structures is response estimation of structures subject to ground motions in multiple intensities. Various simplified procedures for seismic analyses have been used by researchers in order to overcome the huge computational demand involved in assessment of several design alternatives. Nevertheless, cost assessment has been mostly used in comparative study among a limited number of design alternatives, and incorporation of life-cycle cost

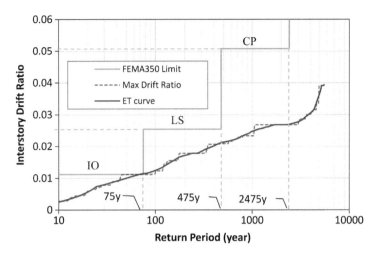

FIGURE 9.9 Performance curve (ET curve) for the performance-based design.

directly in the design process has attracted the attention of researchers (Mitropoulou, Lagaros, and Papadrakakis 2011; Kaveh, Laknejadi, and Alinejad 2012; Frangopol, Strauss, and Bergmeister 2009). Push-over analysis has been widely used as a seismic assessment tool in this area. However, well-known limitations of this analytical tool – besides its disability in properly estimating nonstructural cost components due to floor acceleration – have increased the need for more realistic and reliable dynamic analysis procedures with a tolerable computational demand. In this section, ET analysis has been used to estimate seismic response of the structure, and the procedure to calculate the required cost components has been formulated.

The total cost C_{TOT} of a structure can be considered as the sum of its initial construction cost C_{IN}, which is function of design vector s and the present value of the life-cycle cost C_{LC}, which is function of lifetime t and the design vector s (Mitropoulou, Lagaros, and Papadrakakis 2011):

$$C_{TOT}(t,s) = C_{IN}(s) + C_{LC}(t,s) \tag{9.1}$$

9.6.1 INITIAL COSTS

Initial cost is the construction cost of a new structure or the rehabilitation cost of an existing facility. In our design example, which is a new moment-resisting steel frame, the initial cost is related to the land price, material, and the labor cost for the construction of the building. As the land price and nonstructural components cost are constant for all design alternatives, they can be eliminated from the total cost calculation, and the initial steel weight of the structure with a labor overhead can be considered as representor of the initial cost. So, an initial cost equal to $500 per m² over the 700 m² total area of the structure for the prescriptive design is considered and, for other design alternatives, it will be calculated according to their steel weight difference by a material plus labor cost of 2 $/kg.

9.6.2   Life-Cycle Cost

Life-cycle cost in this study refers to the consequent costs resulting from earthquakes that may occur during the lifetime of the structure. Based on the recent literature, multiple limit states according to interstory drift ratio are considered. These limit states and damages depend on the performance of both structural and nonstructural components. In order to calculate the life-cycle cost of the structure, the following cost components are involved: damage repair cost, cost of loss of contents due to structural damage quantified by the maximum interstory drift and floor acceleration, the loss of rental cost, the loss of income cost, the cost of injuries, and the cost of human fatalities (Mitropoulou, Lagaros, and Papadrakakis 2010; Wen and Kang 2001b).

A correlation is required to quantify these losses in economic terms. Several damage indices have been used to quantify the seismic performance of structures. Commonly, interstory drift (Δ) has been considered as a measure of both structural and non-structural damage. In this study, seven limit states according to drift ratios based on ATC-13 (1985) are used to describe structural performance as shown in Table 9.2. On the other hand, maximum floor acceleration is used to quantify the loss of contents. The relation between floor acceleration values and damage states are shown in Table 9.2, based on a work by Elenas and Meskouris (2001). The addition of the maximum floor acceleration component in life-cycle cost calculation is introduced by Mitropoulou, Lagaros, and Papadrakakis (2010). The piecewise linear relation has been assumed in order to establish a continuous relation between damage indices and costs (Mirzaee and Estekanchi 2013).

Expected annual cost is found to be the most proper intermediate parameter to calculate life-cycle cost of structures using the ET method. The procedure and

---

TABLE 9.2
Drift Ratio and Floor Acceleration Limits for Damage States

Performance level	Damage states	Drift ratio limit (%) ATC-13 (1985)	Floor acceleration limit (g) (Elenas and Meskouris 2001)
I	None	$\Delta \leq 0.2$	$a_{floor} \leq 0.05$
II	Slight	$0.2 < \Delta \leq 0.5$	$0.05 < a_{floor} \leq 0.10$
III	Light	$0.5 < \Delta \leq 0.7$	$0.10 < a_{floor} \leq 0.20$
IV	Moderate	$0.7 < \Delta \leq 1.5$	$0.20 < a_{floor} \leq 0.80$
V	Heavy	$1.5 < \Delta \leq 2.5$	$0.80 < a_{floor} \leq 0.98$
VI	Major	$2.5 < \Delta \leq 5$	$0.98 < a_{floor} \leq 1.25$
VII	Destroyed	$5.0 < \Delta$	$1.25 < a_{floor}$

formulation whose validity is investigated by Kiureghian (2005) to be used in ET framework is described here in detail.

A common framework for performance-based earthquake engineering, used by researchers at the Pacific Earthquake Engineering Research (PEER) Center, can be summarized by Equation (9.2), named as PEER framework formula. By the use of this formula the mean annual rate (or annual frequency) of events (e.g., a performance measure) exceeding a specified threshold can be estimated (Kiureghian 2005).

$$\lambda(dv) = \int_{dm} \int_{edp} \int_{im} G(dv|dm)|dG(dm|edp)||dG(edp|im)||d\lambda(im)| \qquad (9.2)$$

where:

im: an intensity measure (e.g., the peak ground acceleration or spectral intensity)
edp: an engineering demand parameter (e.g., an interstory drift)
dm: a damage measure (e.g., the accumulated plastic rotation at a joint)
dv: a decision variable (e.g., Dollar loss, duration of downtime)

Here $G(x|y) = P(x < X|Y = y)$ denotes the Conditional Complementary Cumulative Distribution Function (CCDF) of random variable X given $Y = y$, and $\lambda(x)$ is the mean rate of $\{x < X\}$ events per year. All of the aleatory and epistemic uncertainties present in describing the model of the structure and its environment, and also the stochastic nature of earthquakes, can be properly modeled in this formula. It should be noted that the deterioration of the structure has been ignored, and it has been assumed that it is instantaneously restored to its original state after each damaging earthquake. Another fundamental assumption is that, conditioned on *EDP*, *DM* is independent of *IM*, and, conditioned on *DM*, *DV* is independent of *EDP* and *IM*. The later assumption makes it possible to decompose the earthquake engineering task into subtasks presented in Figure 9.10. Note that the ET method is used in the response analysis box in this flowchart and will create a proper baseline to calculate the following boxes.

By considering various cost components as decision variable *dv* and using Equation (9.2), $\lambda(dv)$ the annual rate that the cost component values (*DV*) exceeds certain value *dv* can be obtained. Results can be presented by a curve with cost values *dv* in horizontal axis and annual rate of exceedance as vertical axis known as "Loss Curve" (Yang, Moehle, and Stojadinovic 2009).

For a variable X, the differential quantity $|\lambda(x + dx) - \lambda(x)| \cong |d\lambda(x)|$ describes the mean number of events $\{x < X \leq x + dx\}$ per year. Thus, assuming X is non-negative, its expected cumulative value in one year is

$$E[\Sigma X] = \int_0^\infty x|d\lambda(x)| = \int_0^\infty \lambda(x)dx \qquad (9.3)$$

It can be inferred that, the area underneath the $\lambda(x)$ vs. x curve gives the mean cumulative value of X for all earthquake events occurring in one year time. Therefore,

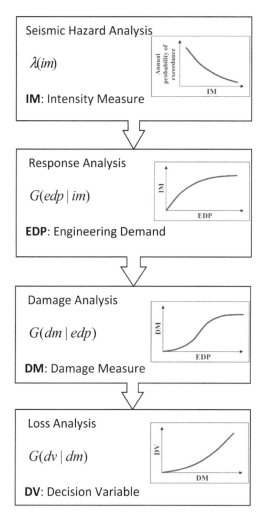

FIGURE 9.10 Performance-based earthquake engineering framework (Yang, Moehle, and Stojadinovic 2009).

in our problem where x is the cost component values as decision variable, the area under $\lambda(dv)$ vs. dv curve (i.e., Loss Curve) represents the mean cumulative annual considered component cost for all earthquake events in one year.

As a practical procedure, Loss Curve can be acquired from ET curve mentioned above. Here, the annual probability of exceedance of drift ratios should be determined. By reversing the return period on the x-axis to obtain the mean annual rate of exceedance and using it on the y-axis, the annual rate of exceedance of the interstory drift can be obtained. If the interstory drift is replaced by component cost applying the linear relationship discussed previously using Table 9.2, the annual rate of exceedance for component can be obtained. This curve is the Loss Curve being

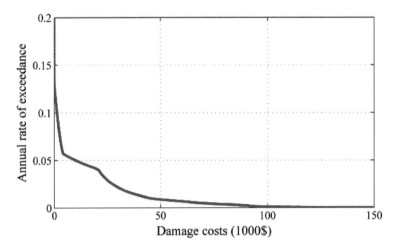

FIGURE 9.11   A sample Loss Curve due to damage cost.

sought. A Loss Curve due to floor acceleration can be easily obtained similarly. In Figure 9.11 a sample loss curve due to damage cost is depicted. The area under the loss curve represents the mean annual component cost caused by all earthquakes in one year.

As it mentioned life-cycle cost consists of several components and can be calculated as follows:

$$C_{LC} = C_{dam} + C_{con} + C_{ren} + C_{inc} + C_{inj} + C_{fat} \tag{9.4}$$

$$C_{con} = C_{con}^{\Delta} + C_{con}^{acc} \tag{9.5}$$

where C_{dam} = the damage repair cost; C_{con}^{Δ} = the loss of contents cost due to structural damage quantified by interstory drift; C_{con}^{acc} = the loss of contents cost due to floor acceleration; C_{ren} = the loss of rental cost; C_{inc} = the cost of income loss; C_{inj} = the cost of injuries and C_{fat} = the cost of human fatality. Formulas to calculate each cost component can be found in Table 9.3. The first term of each formula is presented in the last column of the table as the basic cost. The values of the mean damage index, loss of function, downtime, expected minor injury rate, expected serious injury rate and expected death rate used in this study are based on ATC-13 (1985) restated in FEMA-227 (1992). Table 9.4 provides these parameters for each damage state. Loss of function time and down time are considered as the time required to recover the full functionality of the building based on a table from ATC-13 (1985) for earthquake engineering facility classification 16 and medium rise moment-resisting steel frame. Also, occupancy rate is taken 2 persons per 100 m². Note that these are an estimation of cost components and a detailed assessment is necessary to evaluate the expected cost. The method, with no limitation, has the capability of incorporating detailed calculation on cost components.

TABLE 9.3
Formulas for Cost Components Calculation in Dollars

Cost component	Formula	Basic cost
Damage repair (C_{dam})	Replacement cost × floor area × mean damage index	400 \$/m²
Loss of contents (C_{con})	Unit contents cost × floor area × mean damage index	150 \$/m²
Loss of rental (C_{ren})	Rental rate × gross leasable area × loss of function time	10 \$/month/m²
Loss of income (C_{inc})	Income rate × gross leasable area × down time	300 \$/year/m²
Minor injury ($C_{inj,m}$)	Minor injury cost per person × floor area × occupancy rate × expected minor injury rate	2,000 \$/person
Serious injury ($C_{inj,s}$)	Serious injury cost per person × floor area × occupancy rate × expected serious injury rate	20,000 \$/person
Human fatality (C_{fat})	Human fatality cost per person × floor area × occupancy rate × expected death rate	300,000 \$/person

Source: Mitropoulou, Lagaros, and Papadrakakis 2011; Wen and Kang 2001a; ATC-13 1985.

TABLE 9.4
Damage State Parameters for Cost Calculations

Damage states	Mean damage index (%)	Expected minor injury rate	Expected serious injury rate	Expected death rate	Loss of function time (days)	Down time (days)
(I)-None	0	0	0	0	0	0
(II)-Slight	0.5	0.00003	0.000004	0.000001	1.1	1.1
(III)-Light	5	0.0003	0.00004	0.00001	16.5	16.5
(IV)-Moderate	20	0.003	0.0004	0.0001	111.8	111.8
(V)-Heavy	45	0.03	0.004	0.001	258.2	258.2
(VI)-Major	80	0.3	0.04	0.01	429.1	429.1
(VII)-Destroyed	100	0.4	0.4	0.2	612	612

Source: ATC-13 1985; FEMA-227 1992.

According to Equation (9.1) the total life-cycle cost is considered as the sum of the initial construction costs and the present value of the annual damage costs summed up through the life time of the structure. A discount rate equal to 3 percent over a 50-year life of the building has been considered to transform the damage costs to the present value and calculate the expected damage cost of the structure in its lifetime. This total cost is used as the objective function in optimization algorithm and a design with the lowest total cost is being sought. Due to capabilities of genetic algorithm this design is the global optimum alternative with a high chance.

As the previous sections, genetic algorithm (GA) has been used to find the optimum design. Alternative designs should meet some initial constraints. One of the constraints is strong column and weak beam criterion which should be checked and the other constraint that should be considered before the analysis phase is that the selected sections for columns in each story should not be weaker than the upper story. Beside these constraints, all AISC360 checks must be satisfied for the gravity loads. Once the expressed constraints are satisfied, the LCC (Life Cycle Cost) analysis is performed. It is important to note that each of these feasible organisms is acceptable design according to the code ignoring seismic actions. But, in order to reach the optimum solution, algorithm will reproduce new design alternatives based on the initial population and mutate until the stop criteria is met. The flowchart of the applied methodology is presented in Figure 9.12.

Genetic algorithm with an initial population size of 200 leads to an optimum design after about 2,600 ET response history analyses. Total costs for feasible design alternatives in optimization procedure are depicted in Figure 9.13. The optimum design sections are presented in Figure 9.14 and its performance in various seismic intensities (i.e., ET curve) is presented in Figure 9.15. It can be seen that the structure satisfies performance limitations of FEMA-350 with a margin that is justified by economic concerns.

9.7   COMPARATIVE STUDY

In this section, components of life-cycle cost for the three structures (i.e., prescriptive, performance-based, and value-based designs) are compared. These structures are design optimally based on various design philosophies. In Figure 9.16 cost components for the three structures are provided in $1,000. Each bar presents contribution of various cost components and the value of total cost for each design can be found above the bars. Components in bars are in the same order as the legend. As it can be seen, the prescriptive design has the least initial cost but the largest total cost among three and the value-based design having a larger initial cost has the least total cost in long term. Also, the value-based design has a larger cost of content loss due to floor acceleration. It may reaffirm the sophistications involved in selecting a desired design alternative. In Table 9.5 initial costs based on used initial material, present value of life-cycle costs due to seismic hazards with various exceedance probabilities and the determinative part that is, total cost of three structures are presented. It can be verified that a value-based design has the least total cost and would be an economical alternative in long term. An increase of $12,200 in initial material cost over the prescriptive design will lead to a decrease of $87,400 in expected life-cycle cost having totally $75,200 profit. Although PBD has a minor expected total cost in comparison with the prescriptive design, neither the prescriptive design criteria nor the performance based one will necessarily lead to an economical design in long term.

9.8   SUMMARY AND CONCLUSIONS

A framework for direct use of the concept of value in the structural design procedure incorporating the benefits of ET method has been established. Application of the ET

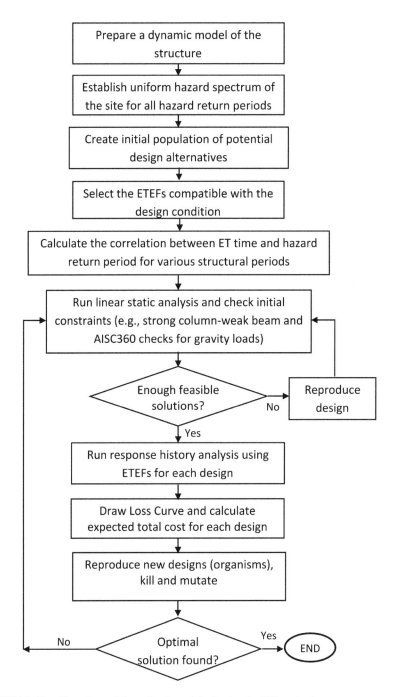

FIGURE 9.12 Flowchart of the value-based design by the ET method.

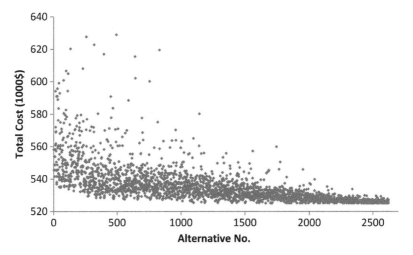

FIGURE 9.13   Total costs for feasible design alternatives in optimization procedure.

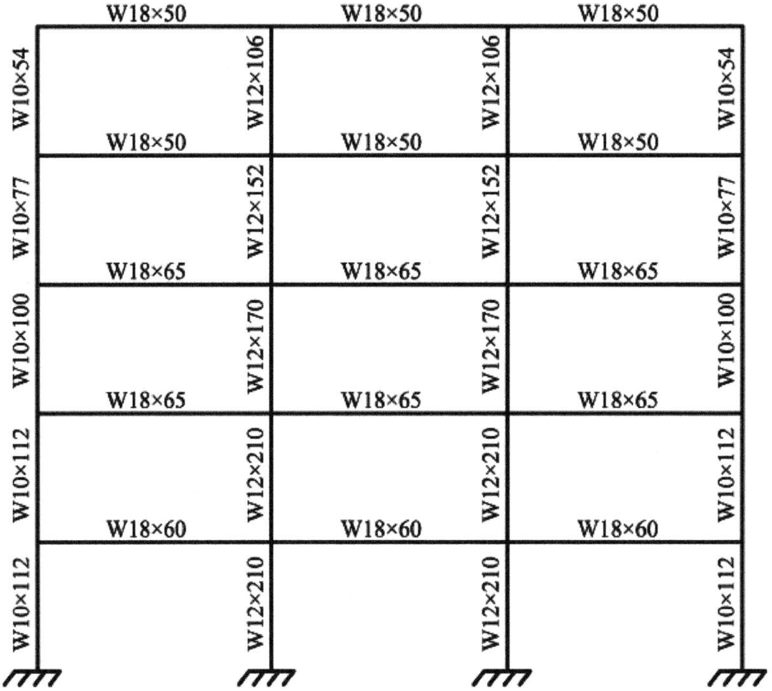

W18×50	W18×50	W18×50

FIGURE 9.14   Value-based design sections of the frame.

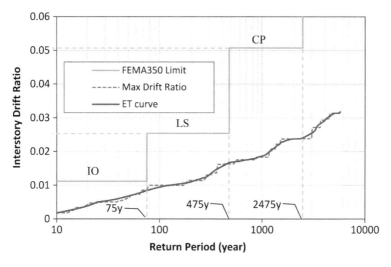

FIGURE 9.15   Performance curve (ET curve) for the value-based design.

TABLE 9.5
Values of Life-cycle Cost Terms for the Three Designs ($1,000)

Design type	Initial cost	Life-cycle cost	Total cost
Prescriptive	350	250.3	600.3
Performance based	351.7	237.9	589.6
Value based	362.2	162.9	525.1

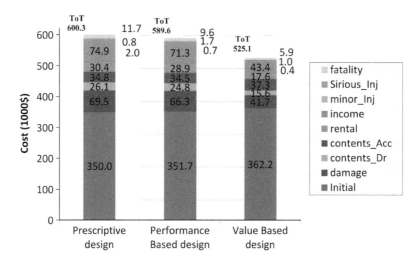

FIGURE 9.16   Cost components and total cost for the three designs ($1,000).

analysis in LCCA has been formulated. ET method and resultant performance curve has provided a proper baseline to calculate expected damage cost, while the required computational effort is in an acceptable range to be used in conventional optimization techniques. To demonstrate the method and compare it with prescriptive and PBD criteria, a five-story moment-resisting frame has been optimally designed according to three distinct design philosophies: a prescriptive design code, a PBD guideline and also the introduced methodology namely value-based design of structures. Structural performance and life-cycle cost components for the three structures have been compared. The resultant prescriptive, performance-based and value-based designs of the frame are different due to their distinct basic design philosophies. Results show that the code-based design of the structure will not necessarily result in an economical design with lower total cost in lifetime of the structure. PBD in this case turns out to require higher initial material cost in comparison with the prescriptive design due to its more restricting limitations, and as expected, better performance in various hazard intensities. The value-based design, however, demands the highest initial material cost, yet the least total cost among three, justifying the increased initial cost. The described methodology provides a pathway toward practical value-based seismic design. It also shows that conventional design procedures based on compliance to design code requirements or performance objectives do not assure achievement of the best final design regarding the overall applicable design values.

9.9 NOMENCLATURE

a_{floor}	Floor acceleration
ATC	Applied technology council
CCDF	Conditional complementary cumulative distribution function
CP	Collapse prevention
C_{con}	Loss of contents cost
C_{con}^{acc}	Loss of contents cost due to floor acceleration
C_{con}^{Δ}	Loss of contents cost due to interstory drift
C_{dam}	Damage repair cost
C_{fat}	Cost of human fatality
C_{inc}	Loss of income cost
C_{inj}	Cost of injuries
$C_{inj,m}$	Cost of minor injuries
$C_{inj,s}$	Cost of serious injuries
C_{ren}	Loss of rental cost
C_{IN}	Initial cost
C_{LC}	Life-cycle cost
C_{TOT}	Total cost
dm	A damage measure threshold
dv	A decision variable threshold
DM	Damage measure
DV	Decision variable
edp	An engineering demand parameter threshold
E	Elastic modulus

EDP	Engineering demand parameter
ET	Endurance time
ETEF	Endurance time excitation function
FEMA	Federal emergency management agency
F_y	Yielding stress
g	Acceleration of gravity
GA	Genetic algorithm
im	An intensity measure threshold
IDA	Incremental dynamic analysis
IM	Intensity measure
IO	Immediate occupancy
INBC	Iranian National Building Code
LCCA	life-cycle cost analysis
LRFD	load resistance factor design
LS	life safety
PBD	Performance-based design
PEER	Pacific earthquake engineering research center
R	Pehavior factor
s	Design vector
t	Lifetime of structure
t_{ET}	ET excitation time
VBD	value-based design
Δ	Interstory drift ratio

NOTE

1 *Chapter Source*: Basim, M.C., H.E. Estekanchi, and A. Vafai. 2016. "A Methodology for Value Based Seismic Design of Structures by the Endurance Time Method." *Scientica Iranica, Transaction A, Civil Engineering* 23, no. 6, p. 2514.

REFERENCES

AISC341. 2010. *Seismic Provisions for Structural Steel Buildings, (ANSI/AISC 341-10)*. Chicago: American Institute of Steel Construction.

AISC360. 2010. *Specification for Structural Steel Buildings (ANSI/AISC 360-10)*. Chicago: American Institute of Steel Construction.

ATC-13. 1985. *Earthquake Damage Evaluation Data for California*. Applied Technology Council.

Elenas, A., and K. Meskouris. 2001. "Correlation Study between Seismic Acceleration Parameters and Damage Indices of Structures." *Engineering Structures* 23, no. 6, pp. 698–704.

Endurance Time Method Website. 2014. online. https://sites.google.com/site/etmethod/.

Estekanchi, H.E., and M.C. Basim. 2011. "Optimal Damper Placement in Steel Frames by the Endurance Time Method." *The Structural Design of Tall and Special Buildings* 20, no. 5, pp. 612–630.

Estekanchi, H., A. Vafai, and M. Sadeghazar. 2004. "Endurance Time Method for Seismic Analysis and Design of Structures." *Scientia Iranica* 11, no. 4, pp. 361–370.

FEMA-227. 1992. *A Benefit–Cost Model for the Seismic Rehabilitation of Buildings.* Washington, DC: Federal Emergency Management Agency, Building Seismic Safety Council.

FEMA-350. 2000. *Recommended Seismic Design Criteria for New Steel Moment-Frame Buildings.* Washington, DC: Federal Emergency Management Agency.

Fragiadakis, M., N.D. Lagaros, and M. Papadrakakis. 2006. "Performance-Based Multiobjective Optimum Design of Steel Structures Considering Life-Cycle Cost." *Structural and Multidisciplinary Optimization* 32, no. 1, pp. 1–11.

Frangopol, D.M., A. Strauss, and K. Bergmeister. 2009. "Lifetime Cost Optimization of Structures by a Combined Condition–Reliability Approach." *Engineering Structures* 31, no. 7, pp. 1572–1580.

Ganzerli, S., C. Pantelides, and L. Reaveley. 2000. "Performance-Based Design Using Structural Optimization." *Earthquake Engineering and Structural Dynamics* 29, no. 11, pp. 1677–1690.

Gupta, A., and H. Krawinkler. 1999. *Seismic Demands for Performance Evaluation of Steel Moment Resisting Frame Structures.* Technical Report 132. John A. Blume Earthquake Engineering Research Center, Department of Civil Engineering, Stanford University.

Hariri-Ardebili, M., S. Sattar, and H. Estekanchi. 2014. "Performance-Based Seismic Assessment of Steel Frames Using Endurance Time Analysis." *Engineering Structures* 69, pp. 216–234.

Ibarra, L.F., R.A. Medina, and H. Krawinkler. 2005. "Hysteretic Models that Incorporate Strength and Stiffness Deterioration." *Earthquake Engineering and Structural Dynamics* 34, no. 12, pp. 1489–1511.

Kappos, A.J., and E. Dimitrakopoulos. 2008. "Feasibility of Pre-Earthquake Strengthening of Buildings Based on Cost-Benefit and Life-Cycle Cost Analysis, with the Aid of Fragility Curves." *Natural Hazards* 45, no. 1, pp. 33–54.

Kaveh, A., K. Laknejadi, and B. Alinejad. 2012. "Performance-Based Multi-Objective Optimization of Large Steel Structures." *Acta Mechanica* 223, no. 2, pp. 355–369.

Kiureghian, A.D. 2005. "Non-Ergodicity and PEER's Framework Formula." *Earthquake Engineering and Structural Dynamics* 34, no. 13, pp. 1643–1652.

Lignos, D.G., and H. Krawinkler. 2010. "Deterioration Modeling of Steel Components in Support of Collapse Prediction of Steel Moment Frames Under Earthquake Loading." *Journal of Structural Engineering* 137, no. 11, pp. 1291–1302.

Liu, Z., S. Atamturktur, and C.H. Juang. 2013. "Performance Based Robust Design Optimization of Steel Moment Resisting Frames." *Journal of Constructional Steel Research* 89, pp. 165–174.

Liu, M., S.A. Burns, and Y. Wen. 2003. "Optimal Seismic Design of Steel Frame Buildings Based on Life Cycle Cost Considerations." *Earthquake Engineering and Structural Dynamics* 32, no. 9, pp. 1313–1332.

Liu, M., S.A. Burns, and Y. Wen. 2005. "Multiobjective Optimization for Performance-Based Seismic Design of Steel Moment Frame Structures." *Earthquake Engineering and Structural Dynamics* 34, no. 3, pp. 289–306.

Mazzoni, S., F. McKenna, M.H. Scott, G.L. Fenves, and B. Jeremic. 2006. *Open System for Earthquake Engineering Simulation (OpenSees).* Berkeley: University of California..

Mirfarhadi, S.A., and H.E. Estekanchi. 2020. "Value-based Seismic Design of Structures Using Performance Assessment by the Endurance Time Method." *Structure and Infrastructure Engineering*, 16 no. 10, pp. 1397–1415.

Mirzaee, A., and H.E. Estekanchi. 2013. "Performance-Based Seismic Retrofitting of Steel Frames by Endurance Time Method." *Earthquake Spectra.* Online. http://dx.doi.org/ 10.1193/081312EQS262M.

Mirzaee, A., H. Estekanchi, and A. Vafai. 2010. "Application of Endurance Time Method in Performance-Based Design of Steel Moment Frames." *Scientia Iranica* 17, pp. 361–370.

Mirzaee, A., H. Estekanchi, and A. Vafai. 2012. "Improved Methodology for Endurance Time Analysis: from Time to Seismic Hazard Return Period." *Scientia Iranica* 19, no. 5, pp. 1180–1187.

Mitropoulou, C.C., N.D. Lagaros, and M. Papadrakakis. 2010. "Building Design Based on Energy Dissipation: A Critical Assessment." *Bulletin of Earthquake Engineering* 8, no. 6, pp. 1375–1396.

Mitropoulou, C.C., N.D. Lagaros, and M. Papadrakakis. 2011. "Life-Cycle Cost Assessment of Optimally Designed Reinforced Concrete Buildings Under Seismic Actions." *Reliability Engineering & System Safety* 96, no. 10, pp. 1311–1331.

Nozari, A., and H. Estekanchi. 2011. "Optimization of Endurance Time Acceleration Functions for Seismic Assessment of Structures." *International Journal of Optimization in Civil Engineering* 1, no. 2, pp. 257–277.

Pan, P., M. Ohsaki, and T. Kinoshita. 2007. "Constraint Approach to Performance-Based Design of Steel Moment-Resisting Frames." *Engineering Structures* 29, no. 2, pp. 186–194.

Takahashi, Y., A.D. Kiureghian, and A.H.S. Ang. 2004. "Life-Cycle Cost Analysis Based on a Renewal Model of Earthquake Occurrences." *Earthquake Engineering and Structural Dynamics* 33, no. 7, pp. 859–880.

Wen, Y., and Y. Kang. 2001a. "Minimum Building Life-Cycle Cost Design Criteria. I: Methodology." *Journal of Structural Engineering* 127, no. 3, pp. 330–337.

Wen, Y., and Y. Kang. 2001b. "Minimum Building Life-Cycle Cost Design Criteria. II: Applications." *Journal of Structural Engineering* 127, no. 3, pp. 338–346.

Yang, T.Y., J.P. Moehle, and B. Stojadinovic. 2009. *Performance Evaluation of Innovative Steel Braced Frames*. University of California, Berkeley.

10 Seismic Resilient Design by ET

10.1 INTRODUCTION

The resilience in cities and accompanying optimum allocation of public resources necessitates structures with predictable and reliable performance in the case of natural hazards.[1] Earthquakes are considered one of the most destructive and costly natural hazards that threaten cities in seismically active regions. So, assessment of seismic safety and performance of buildings and structural components are among the major challenges in Earthquake Engineering. Reliability and accuracy of seismic analysis procedure is a key concern in almost all seismic-assessment procedures for both new and existing structures, especially in modern approaches to seismic design. Various limitations of simplified seismic analyses have increased the need for more realistic and reliable dynamic analysis procedures. The Endurance Time (ET) method is a response-history based seismic assessment procedure whereby structures are subjected to gradually intensifying dynamic excitations, and their performance is evaluated based on their response at different excitation levels correlated to specific ground-motion intensities (Estekanchi, Valamanesh, and Vafai 2007). This procedure considerably reduces the required huge computational demand of a complete response-history analysis while maintaining the major benefits of it – that is, accuracy and insensitivity to model complexity. These viable advantages provide the prerequisites to directly incorporate the new age design concerns such as life-cycle cost of the structure or resiliency measures in design procedure (Basim and Estekanchi 2015). The main objective in this chapter is to explore the use of the ET method in evaluating seismic resiliency of a construction in quantitative terms.

The concept of disaster resilience in communities has developed rapidly in recent years. The need to emphasize the preparedness of communities to recover from disasters has been confirmed in the 2005 World Conference on Disaster Reduction (WCDR). The aim is to be prepared and to be able to recover in an acceptable time from an unexpected shock in the community and, meanwhile, reduce its vulnerability. The overview of intuitive definitions of resiliency can be found on a work by Manyena (2006). Some frameworks are introduced to provide quantitative evaluation of resilience. These methods can be considered as complementary analysis beyond estimating losses. Resilience measures should take into account technical, social, and economic impacts of a disaster to cover the vast definition of resilience (Cimellaro, Reinhorn, and Bruneau 2010). A general framework for evaluating community resilience has been introduced by Bruneau et al. (2003). They used complementary measures of resilience as reduced failure probabilities, reduced consequences from failures, and reduced time to recovery. They used four dimensions of resiliency for a system: robustness, rapidity, resourcefulness, and redundancy. Chang and Shinozuka (2004) also introduced a measure of resilience that relates expected losses in future disasters to a community's seismic performance objectives and implemented the method in a case study of the Memphis, Tennessee, water delivery system.

Many uncertain parameters are involved in resilience of a construction in the case of a natural or man-made hazard. Bruneau and Reinhorn (2007) tried to relate probability functions, fragilities, and resilience in a single integrated approach for acute-care facilities. Cimellaro, Reinhorn, and Bruneau (2010) proposed a framework to evaluate disaster resilience based on dimensionless analytical functions related to the variation of functionality during a period of interest, including losses in the disaster and in the recovery path. Losses are described as functions of fragility of systems that are determined using multidimensional performance limit thresholds to account for uncertainties. They implemented the method for a typical California hospital building and a hospital network, considering direct and indirect losses. Their proposed framework, with some modifications, is the underlying basis of the present study.

Value-based seismic design of structures using the ET method has been introduced in a work by Basim and Estekanchi (2014). In this methodology, Life-Cycle Cost Analysis (LCCA) has been used in order to evaluate the performance of the structure during its life span in economic terms. This analysis can provide a baseline to incorporate technical, economic, and social (or any other) intended measures thought to be impressive in the resilience of cities in design procedure. The broad concept of resilience demands a flexible design framework to employ these several criteria from various fields of expertise in the design stage. LCCA demands performance assessment of the structure in multiple hazard levels. Considering the required repetitive and massive analyses in this procedure, the application of the ET method in combination with the concept of LCCA can provide the means to use economic concerns directly in the design stage.

In order to demonstrate the proposed method of quantitative evaluation of resilience by the ET method, a prototype structure of a hospital building located in Tehran is considered. It is first optimally designed according to the Iranian National Building Code (INBC). Then, FEMA-350 (2000) limitations as a performance-based design criteria are applied and, finally, new design sections are acquired through the value-based design method. In the third design approach, the intention is to design a structure for the minimum total cost during its life time. The resilience of the three different designs is evaluated using the proposed method, and results are compared and discussed. Reduced computational demand in the ET analysis method provides the prerequisites to use optimization algorithms in design procedure. Although resiliency measures are not directly incorporated in optimization procedure here, this work is intended to pave the way toward the practical design of construction with improved resiliency.

10.2   EARTHQUAKES AND RESILIENCY

Cities cannot be considered resilient if they are not protected against the dangers and potential damage that may be imposed by natural hazards. Earthquakes are considered one of the most destructive and costly natural hazards that threaten cities. So, stability of the community during and after seismic hazards is thought to have a determinative impact on the resilience of cities in seismically active regions. Resilience may have broad measures in the whole city as a body, or sub-measures in individual buildings. Also, the impact of seismic hazards on a community may be studied from various points of view, and various concepts may be defined as resilience – such as time to recovery, life safety, or damage reduction. For example, besides life safety, "downtime" seems to be an impressive resilience measure for a hospital building or a fire station, and it is wise to consider these measures with a reasonable portion in the design stage. Some limitations may be required for such critical facilities, too.

Incorporation of seismic resilience factors in design procedure requires mitigation from common design procedures intended to focus on a limited number of objectives, such as structural performance or loss prevention to a broader one with the capability to incorporate any desired and advancing terms in priority measures among design alternatives. Codes for building design, commonly set some minimum compliance-based standards and, in performance terms, we can be confident that they will provide safe buildings, but they promise little in terms of recovery. Readily introduced methodology can provide a wider description of the design target by defining the earthquake consequences: structural damage, loss of contents, losses due to downtime, human injuries, and fatalities in the form of quantifiable parameters. In this way, it is expected that the resultant design will perform with desired post-earthquake capabilities with manageable disruption.

10.3   CONCEPTS OF ENDURANCE TIME METHOD

A reliable estimation of the damage to various structures and their compartments requires realistic evaluation of seismic response of structures when subjected to strong ground motions. This in turn requires the development and utilization of advanced numerical techniques using reasonably realistic dynamic modeling. While any serious development in the area of seismic-resistant design has to be backed up with decent real-world experimental investigation: the type and number of decision variables (DVs) are usually so diverse that numerical investigations remain the only practical alternative in order to seek good solutions regarding performance and safety.

In the ET method, structures are subjected to a predesigned intensifying dynamic excitation, and their performance is monitored continuously as the level of excitation is increased (Estekanchi, Vafai, and Sadeghazar 2004). A typical ET Excitation Function (ETEF) is shown in Figure 10.1. The level of excitation, or excitation intensity, can be selected to be any relevant intensity parameter considering the nature of the structure or component being investigated.

Classically, parameters such as peak ground acceleration (PGA) or spectral intensity have been considered the most relevant intensity parameters in structural design.

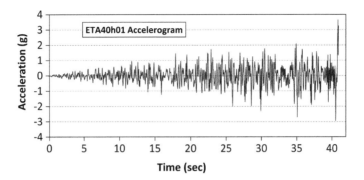

FIGURE 10.1   Typical ET record incorporating intensifying dynamic excitation.

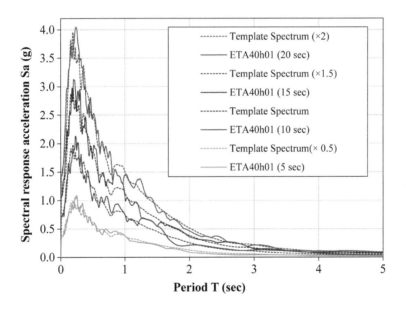

FIGURE 10.2   Typical response spectra of ET records at various times (ETA40h01).

More recently, parameters based on input energy, displacement, and damage spectra are being proposed as a better representative of the dynamic excitation intensity considering structural response. Figure 10.2 shows the response spectra produced by a typical ETEF at various times. A number of ETEFs are publicly available through the ET method website (Estekanchi 2014).

While response spectra have been considered a standard measure of intensity in producing currently available ETEFs, other intensity measures can also be considered as well. As can be expected, most of these intensity measures are correlated to each other, and the challenge is to choose a best combination of various parameters to achieve better intensifying excitations that can produce better output. Here, the response spectra have been considered as the intensity parameter, and ETEF has been produced in such a way that the response spectra produced by each window from time 0 to t is proportional to a spectrum.

The application of the ET method in performance-based design was studied by Mirzaee, Estekanchi, and Vafai (2010), introducing the "ET curve" and the "Target Curve," which respectively express the seismic performance of a structure along various seismic intensities and their limiting values according to code recommendations. Substituting return period or annual probability of exceedance for time in the expression of the performance will make the presentation of the results more explicit, and their convenience for calculating probabilistic cost will be increased (Mirzaee, Estekanchi, and Vafai 2012). Also, damage levels have been introduced to express the desired damage states in quantifiable terms.

The hazard return period corresponding to a particular time in ET analysis can be calculated by matching the response spectra at effective periods – for example, from 0.2 to 1.5 times of the structure's fundamental period of vibration. The procedure is based on the coincidence of response spectra obtained from the ET accelerogram at different times, and response spectra defined for Tehran at different hazard levels. In Figure 10.3 a sample target curve and ET curve considering various performance criteria are depicted where ET analysis time has been mapped into a return period on a horizontal axis. As can be seen, the structure satisfies the code IO (Immediate Occupancy) level limitations, but it has violated the LS (Life Safety) and CP (Collapse Prevention) levels limitations, and the frame does not have acceptable performance. It can be inferred that one of advantages of the ET method is that the performance of the structure in continuous increasing hazard levels can be properly depicted in an easy-to-read figure.

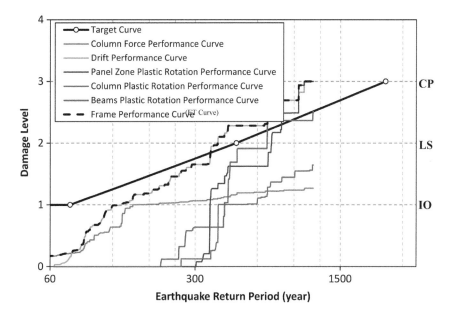

FIGURE 10.3 Performance assessment by ET method.

10.4   VALUE-BASED SEISMIC DESIGN BY THE ET METHOD

Life-cycle cost analysis (LCCA) has become an important part of structural engineering to assess the structural comeback and evaluate the performance of the structure in economic terms. It has gained considerable attention from decision-making centers deciding on the most cost-effective solution related to the construction of structures in seismic regions. LCCA has provided a reliable tool for estimating damage costs due to future earthquakes during the design life of a structure. Instead of "cost" in dollars, in the decision-making process any other measure can be used to compare and evaluate a design alternatives' expected operation. In this section, the total expected cost imposed by earthquake occurrences during a structure's life span is selected as an evaluation measure, since engineers might be more familiar with this concept. A correlation will be required to express other measures mentioned above, such as downtime or social impacts in economic terms and in dollars. By the use of this method the expected total cost of a structure, including the initial cost and also losses resulting from earthquakes during its life span, can be considered as the main indicator of the priority of design alternatives. This analysis in tandem with an optimization algorithm can result in a design offering the lowest total cost. LCCA demands the calculation of the cost components that are related to the performance of the structure in multiple earthquake hazard levels. However, these calculations require repetitive and massive analyses of performance assessment and huge computational demand, and the sophistication involved may make optimization algorithms impractical or the simplifications used may decrease the reliability of the outcome. Application of the ET method in combination with the concept of LCCA has led to development of a framework for practical Value-based Seismic Design of structures (Mirfarhadi and Esteknchi 2020).

ET analysis provides a proper baseline to perform economic analyses on design alternatives with acceptable computational cost. While value can be defined and considered in its broad sense for design purposes, for the clarity of explanation, the structure that is more economical to construct and maintain is considered the most valued. Initial construction costs and expected seismic damage costs throughout the life time of the structure are usually the two most important parameters for decision-making (Mitropoulou, Lagaros, and Papadrakakis 2011). The cost model used in the present study can be found in detail in a work by Basim and Estekanchi (2014). In this model the total cost C_{TOT} of a structure can be considered as the sum of its initial construction cost C_{IN}, which is a function of the design vector s and the present value of the life-cycle cost C_{LC} which is a function of life time t and the design vector s (Mitropoulou, Lagaros, and Papadrakakis 2011).

Initial cost is the construction and equipping cost of a structure. In our hospital building design example, which is a new moment resisting steel frame, the initial cost is related to the land price, material, and the labor cost for the construction of the structure and equipping costs for health-care facilities. The land price and non-structural components cost are constant for all design alternatives.

To calculate the life-cycle cost of the structure, the following cost components are involved: the damage repair cost, the cost of loss of contents due to structural damage

TABLE 10.1
Drift Ratio and Floor Acceleration Limits for Damage States

Performance level	Damage states	Drift ratio limit (%) ATC-13 (1985)	Floor acceleration limit (g) Elenas and Meskouris (2001)
I	None	$\Delta \leq 0.2$	$a_{floor} \leq 0.05$
II	Slight	$0.2 < \Delta \leq 0.5$	$0.05 < a_{floor} \leq 0.10$
III	Light	$0.5 < \Delta \leq 0.7$	$0.10 < a_{floor} \leq 0.20$
IV	Moderate	$0.7 < \Delta \leq 1.5$	$0.20 < a_{floor} \leq 0.80$
V	Heavy	$1.5 < \Delta \leq 2.5$	$0.80 < a_{floor} \leq 0.98$
VI	Major	$2.5 < \Delta \leq 0.5$	$0.98 < a_{floor} \leq 1.25$
VII	Destroyed	$5.0 < \Delta$	$1.25 < a_{floor}$

quantified by the maximum interstory drift and also floor acceleration, the loss of rental cost, the loss of income cost, the cost of injuries, and the cost of human fatalities (Wen and Kang 2001; Mitropoulou, Lagaros, and Papadrakakis 2010). Interstory drift (Δ) has been considered a measure of both structural and nonstructural damage, and maximum floor acceleration is used to quantify the loss of contents. In this study, seven limit states according to drift ratios based on ATC-13 (1985) are used to describe structural performance as shown in Table 10.1. The relation between floor acceleration values and damage states is shown in Table 10.1, based on a work by Elenas and Meskouris (2001). The addition of the maximum floor acceleration component in life-cycle cost calculation is introduced by Mitropoulou, Lagaros, and Papadrakakis (2010). A piecewise linear relation has been assumed between damage indices and costs (Mirzaee and Estekanchi 2015).

Life-cycle cost of the structure is calculated by summing the cost components as follows:

$$C_{LC} = C_{dam} + C_{con} + C_{ren} + C_{inc} + C_{inj} + C_{fat} \tag{10.1}$$

$$C_{con} = C_{con}^{\Delta} + C_{con}^{acc} \tag{10.2}$$

where C_{dam} = the damage repair cost; C_{con}^{Δ} = the loss of contents cost due to structural damage quantified by interstory drift; C_{con}^{acc} = the loss of contents cost due to floor acceleration; C_{ren} = the loss of rental cost; C_{inc} = the cost of income loss; C_{inj} = the cost of injuries; and C_{fat} = the cost of human fatality. Formulas to calculate each cost component are depicted in Table 10.2. The values of the mean damage index, loss of function, downtime, expected minor injury rate, expected serious injury rate and

TABLE 10.2
Formulas for Cost Components Calculation in Dollars

Cost component	Formula	Basic cost
Damage repair (C_{dam})	Replacement cost × floor area × mean damage index	500 $/m²
Loss of contents (C_{con})	Unit contents cost × floor area × mean damage index	250 $/m²
Loss of rental (C_{ren})	Rental rate × gross leasable area × loss of function time	20 $/month/m²
Loss of income (C_{inc})	Income rate × gross leasable area × down time	300 $/year/m²
Minor injury ($C_{inj,m}$)	Minor injury cost per person × floor area × occupancy rate × expected minor injury rate	2,000 $/person
Serious injury ($C_{inj,s}$)	Serious injury cost per person × floor area × occupancy rate × expected serious injury rate	20,000 $/person
Human fatality (C_{fat})	Human fatality cost per person × floor area × occupancy rate × expected death rate	300,000 $/person

Source: ATC-13 1985, Wen and Kang 2001; Mitropoulou, Lagaros, and Papadrakakis 2011.

TABLE 10.3
Damage State Parameters for Cost Calculations

Damage states	Mean damage index (%)	Expected minor injury rate	Expected serious injury rate	Expected death rate	Loss of function time (days)	Down time (days)
(I)-None	0	0	0	0	0	0
(II)-Slight	0.5	0.00003	0.000004	0.000001	1.1	1.1
(III)-Light	5	0.0003	0.00004	0.00001	16.5	16.5
(IV)-Moderate	20	0.003	0.0004	0.0001	111.8	111.8
(V)-Heavy	45	0.03	0.004	0.001	258.2	258.2
(VI)-Major	80	0.3	0.04	0.01	429.1	429.1
(VII)-Destroyed	100	0.4	0.4	0.2	612	612

Source: ATC-13 1985; FEMA-227 1992.

expected death rate used in this study are based on ATC-13 (1985) restated in FEMA-227 (1992). Table 10.3 provides these parameters for each damage state.

As described in (Basim and Estekanchi 2014), the annual rate by which any cost component exceeds a threshold value is calculated using the PEER framework. This will result in a curve with cost values in the horizontal axis and annual rate of exceedance in the vertical axis known as the "Loss Curve" (Yang, Moehle, and Stojadinovic 2009). In Figure 10.4 a sample loss curve due to damage cost is depicted.

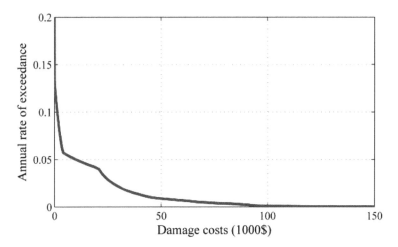

FIGURE 10.4   A typical Loss Curve for the three-story frame.

The area under the loss curve represents the mean annual component cost caused by all earthquakes in one year. The life-cycle cost of the building is the present value of the annual damage costs summed up through the lifetime of the structure. A discount rate equal to 3 percent over a 50-year life of the building has been considered to transform the damage costs to the present value. The total cost of the structure is calculated by summing the initial cost and the life-cycle cost and is used as the objective function in the optimization algorithm seeking a design with the least total cost.

10.5   CASE STUDY: THREE-STORY STEEL MOMENT FRAME

In order to demonstrate the method a three-story and one-bay steel special moment frame used as a hospital building is optimally designed according to the Iranian National Building Code (INBC), which is almost identical to the ANSI/AISC360 (2010) LRFD design recommendations. Also, the frame is designed optimally to conform to FEMA-350 (2000) limitations as a performance-based design criterion and, as a third step, a new design section has been acquired through the value-based design method to have the minimum total cost during its lifetime that is assumed to be 50 years. The performance of the designed frames is investigated by the ET method. For the value-based design, the total cost of the structure is selected as the optimization objective to be minimized. An initial cost equal to $500 per m^2 over the 300 m^2 total area of the structure for the prescriptive design is considered, and for other design alternatives, it will be calculated according to their steel weight difference by a material plus labor cost of 2 $/kg. Occupancy rate is 10 persons per 100 m^2.

Structural response-history analyses were performed in OpenSees (Mazzoni et al. 2006). Genetic algorithm (GA) has been used to find the optimum design. Alternative designs should meet some initial constraints. One of the constraints is strong column and weak beam criterion, which should be checked, and the other constraint that should be considered before the analysis phase is that the selected sections

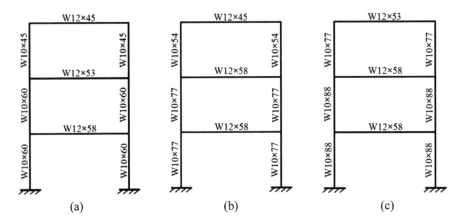

FIGURE 10.5 Frame design results after optimization: (a) codified design (b) performance-based design (c) value-based design (least LCC).

for columns in each story should not be weaker than the upper story. Beside these constraints, all AISC 360 checks must be satisfied for the gravity loads. Once the expressed constraints are satisfied, the LCC (Life Cycle Cost) analysis is performed. A genetic algorithm with an initial population size of 100 leads to an optimum design after about 1,800 ET response-history analyses.

The resultant prescriptive, performance-based and value-based designs of the frame are different due to their distinct basic design philosophies. Design sections for each method are depicted in Figure 10.5. Seismic performance of each design of the frame is shown in Figure 10.6. It can be seen that, for the value-based design, the structure satisfies performance limitations of FEMA350 with a margin that is justified by economic concerns.

10.6   QUANTIFICATION OF RESILIENCE

Resilience can be quantified using a function that presents the ability of the system to sustain its functionality over a period of time. Such a function for a system that has exposed an external shock is presented in Figure 10.7. The system can be a building, infrastructure, lifeline networks, or a whole community. In this figure, the normalized functionality $Q(t)$ of the system is traced during a control time T_{LC} that may be the lifetime of a construction. It is assumed that a disastrous event occurs at a time t_{OE} and it takes a period of time T_{RE} as recovery time in which the system regains its full functionality. Although the final functionality of the system may differ from the initial, it is assumed here that the recovery process restores the under-study building to its initial condition. For a construction under seismic hazards, T_{RE} depends on many external parameters such as hazard intensities, induced damages, management quality, and resources to repair damages. Many uncertainties are involved in the required recovery time and, also, the amount of loss of functionality in the case of an event. The recovery time is known to be the most difficult quantity to predict in this function.

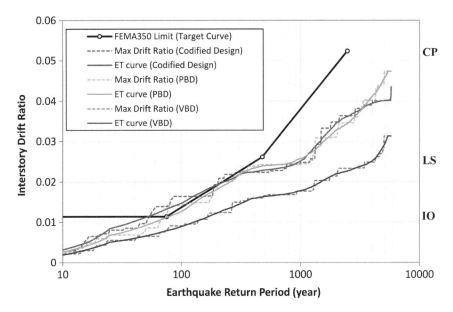

FIGURE 10.6   Comparison of the frames response at various hazard levels.

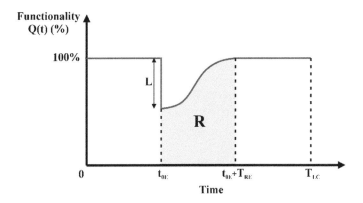

FIGURE 10.7   Schematics of a functionality function for a system.

A resilience measure should represent all the dimensions of resilience, which includes the amount of direct and indirect losses and also the rapidity of the recovery process. According to MCEER (Multidisciplinary Center of Earthquake Engineering to Extreme Event) terminology, resilience is quantified as the area under the functionality curve $Q(t)$ of a system. This measure can be considered as a somewhat comprehensive DV to evaluate the performance of a construction (Bruneau and Reinhorn 2007; Cimellaro, Reinhorn, and Bruneau 2010). Resilience R can be expressed by the following formula as a dimensionless parameter in percentage (Cimellaro, Reinhorn, and Bruneau 2010):

$$R = \int_{t_{0E}}^{t_{0E}+T_{RE}} Q(t) / T_{RE} dt \qquad (10.3)$$

$$Q(t) = 1 - L(1, T_{RE}) \left[H(t - t_{0E}) - H(t - (t_{0E} + T_{RE})) \right] \times f_{Rec}(t, t_{0E}, T_{RE}) \quad (10.4)$$

$L(1, T_{RE})$ is the loss function, and $f_{Rec}(t, t_{0E}, T_{RE})$ is the recovery function; H is the Heaviside step function. Other parameters used have been defined previously. Rapidity in the recovery process can be represented by the slope of the functionality curve ($dQ(t)/dt$). The amount of resources and the quality of management and many other parameters will affect the shape and the slope of the recovery curve and the recovery time T_{RE}. The other dimension of resilience thought to be important in recover capacity of a system is robustness. Usually taken as the residual functionality after a disastrous event and, in the framework discussed by Cimellaro, Reinhorn, and Bruneau (2010), is considered as $1 - \bar{L}(m_L, \alpha\sigma_L)$ where $\bar{L}$ is a random variable with the mean m_L and the standard deviation σ_L and α is a multiplier of the standard deviation corresponding to a specific level of losses. Here, for simplicity of the representation, the estimation of the resilience R is based on the mean values of L. Uncertainties can be modeled using a Monte Carlo approach or reliability methods.

The loss model used in this section is similar to that of the previous section. Of course, many uncertainties are involved in these losses, and various probabilistic loss-estimation methods are proposed in the literature. For simplicity of presentation, a somewhat simple loss model with limited uncertainty calculations is used here. The method has the capability to use more detailed loss-estimation techniques. Total loss L in this framework can be considered as a function of earthquake intensity I and recovery time T_{RE} as it contains both direct losses (L_D) and indirect losses (L_I). The later losses can be directly affected by the recovery time. Each of direct and indirect losses has two subcategories: economic losses and casualties' losses. Therefore, total loss $L(I, T_{RE})$ consists of four contributions: direct economic losses, L_{DE}, direct casualties losses, L_{DC}, indirect economic losses, L_{IE}, and indirect casualties' losses, L_{IC}. In this context, direct economic losses L_{DE} is considered as the sum of damage-repair cost and the loss of contents cost as a ratio of the total building replacement cost. So, L_{DE} is a function of intensity I. More detailed loss models using fragilities can be employed by the formulation presented in the work by Cimellaro, Reinhorn, and Bruneau (2010).

Direct casualties' losses L_{DC} are calculated as a ratio of the instantaneous number of injured or dead people N_{in} to the total number of occupants N_{tot}. This parameter can also be calculated using the model defined in the previous section:

$$L_{DC}(I) = \frac{N_{in}}{N_{tot}} \qquad (10.5)$$

The indirect economic losses may be significant for lifeline systems or any critical facilities such as health-care centers. The indirect economic losses $L_{IE}(I, T_{RE})$

are related to hazard intensity as is also the recovery time. More comprehensive models are required to estimate the post-earthquake losses. Loss of rental and loss of income costs in the used cost model can be considered as components of L_{IE}. Some other components may be involved in lifeline systems such as water or gas delivery networks that may be much more than direct economic losses.

The indirect casualties' losses $L_{IC}(I, T_{RE})$ may be significant for a health-care center. These losses are caused by hospital dysfunction in recovery time after an earthquake. In this framework L_{IC} can be calculated as the ratio of the number of injured persons N_{in} to the total population N_{tot} served or supposed to be served.

$$L_{IC}\left(I, T_{RE}\right) = \frac{N_{in}}{N_{tot}} \tag{10.6}$$

Casualties' losses will affect the total loss as a penalty function using weighting factors according to the following formulas:

$$L_D = L_{DE} \cdot \left(1 + \alpha_{DC} L_{DC}\right)$$

$$L_I = L_{IE} \cdot \left(1 + \alpha_{IC} L_{IC}\right) \tag{10.7}$$

α_{DC}, α_{IC} are the weighting factors representative for the importance of the occupancy that are determined based on social concerns. The total losses L are a combination of direct losses L_D and indirect losses L_I:

$$L\left(I, T_{RE}\right) = L_D\left(I\right) + \alpha_I L_I\left(I, T_{RE}\right) \tag{10.8}$$

α_I is used as a weighting factor to represent the importance of indirect losses to other facilities in a community. It is obvious that the longer recovery time T_{RE} results in higher total-loss values. The next step to calculate resilience of the construction is to estimate a recovery path through which the building regains its functionality. This process is complex and is influenced by many environmental conditions, such as quality of management and amount of resources and may be affected by the amount of disaster consequences in other sectors of the community. The recovery model used in this section is based on the simplified model introduced by Cimellaro, Reinhorn, and Bruneau (2010) with some modifications. A trigonometric function is selected according to the condition experienced in Iran:

$$f_{rec}(t) = a/2\left\{1 + cos\left[\pi b\left(t - t_{0E}\right)/T_{RE}\right]\right\} \tag{10.9}$$

where a, b are constant values that are assumed to be equal to unity (i.e. $a=b=1$). This function is used when the process of recovery starts with considerable delay due

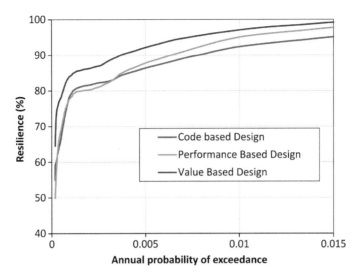

FIGURE 10.8   Resilience curve for the three structures.

to lack of resources or proper management. It is assumed that the structural response to a specific intensity level does not vary in the lifetime of the structure. In other words, deterioration of the structural system is ignored for the sake of simplicity. As noted, the structural responses for any hazard intensity are provided through ET analysis. The results are represented via ET analysis curves. The resilience of the understudy building in case of hazards with any intensity can be calculated using the presented method.

The resilience R of each structure, conditional on the occurrence of earthquakes with any intensity, is depicted in Figure 10.8. In this figure, a vertical axis shows the expected resilience of the structure conditional on the occurrence of an earthquake with the annual probability of exceedance presented in horizontal axis. Using the ET method, the resilience of the structure can be explored in a continuous range of intensities. It is obvious that the value-based design is the most resilient structure among the three alternatives. The reduced computational cost in this framework provides the means to incorporate reliability analyses and account for uncertainties.

10.7   SUMMARY AND CONCLUSIONS

A framework is proposed to calculate a resilience measure using the ET method. A simplified cost and recovery model was developed for a prototype hospital building. Three optimum design alternatives for the structure were considered according to a prescriptive design code: a performance-based design guideline, and a value-based approach. Application of the ET analysis in LCCA has been briefly explained. The ET method and resultant performance curve have provided a proper baseline to calculate expected damage cost, while the required computational effort is in an acceptable range to be used in conventional optimization techniques. Structural performance for the three structures has been compared using the ET curve. A resilience measure according to the current literature was defined and a method to calculate the resilience of the structure conditional on the occurrence of hazard with any intensity was developed. Results were presented in an easy-to-read figure introducing the "Resilience Curve." Results show that the value-based design will have better performance regarding the resilience measures. These curves can be used to compare relative resilience of different designs and thus can provide a basis for achieving the desired seismic resilience through successive design improvements. Although the involved uncertainties were not highlighted in this study, the method has the capability to account for them, requiring a reasonable amount of computational effort.

NOTE

1 *Chapter Source*: Estekanchi, H.E., A. Vafai, and C.B. Mohammad. 2016. "Design and Assessment of Seismic Resilient Structures by the Endurance Time Method." *Scientica Iranica* 23, no. 4, pp. 1648–1657.

REFERENCES

AISC360. 2010. *Specification for Structural Steel Buildings (ANSI/AISC 360-10)*. Chicago: American Institute of Steel Construction.

ATC-13. 1985. *Earthquake Damage Evaluation Data for California*. Redwood City: Applied Technology Council.

Basim, M.C., and H.E. Estekanchi. 2015. "Application of Endurance Time Method in Performance-Based Optimum Design of Structures." *Structural Safety* 56, pp. 52–67.

Bruneau, M., S.E. Chang, R.T. Eguchi, G.C. Lee, T.D. O'Rourke, A.M. Reinhorn, M. Shinozuka, K. Tierney, W.A. Wallace, and D. von Winterfeldt. 2003. "A Framework to Quantitatively Assess and Enhance the Seismic Resilience of Communities." *Earthquake Spectra* 19, no. 4, pp. 733–752.

Bruneau, M., and A. Reinhorn. 2007. "Exploring the Concept of Seismic Resilience for Acute Care Facilities." *Earthquake Spectra* 23, no. 1, pp. 41–62.

Chang, S.E., and M. Shinozuka. 2004. "Measuring Improvements in the Disaster Resilience of Communities." *Earthquake Spectra* 20, no. 3, pp. 739–755.

Cimellaro, G.P., A.M. Reinhorn, and M. Bruneau. 2010. "Framework for Analytical Quantification of Disaster Resilience." *Engineering Structures* 32, no. 11, pp. 3639–3649.

Elenas, A., and K. Meskouris. 2001. "Correlation Study between Seismic Acceleration Parameters and Damage Indices of Structures." *Engineering Structures* 23, no. 6, pp. 698–704.

Estekanchi, H.E. 2014. *Endurance Time Method Website*. https://sites.google.com/site/etmethod/.

Estekanchi, H.E., A. Vafai, and M. Sadeghazar. 2004. "Endurance Time Method for Seismic Analysis and Design of Structures." *Scientia Iranica* 11, no. 4, pp. 361–370.

Estekanchi, H., V. Valamanesh, and A. Vafai. 2007. "Application of Endurance Time Method in Linear Seismic Analysis." *Engineering Structures* 29, no. 10, pp. 2551–2562.

FEMA-227. 1992. *A Benefit–Cost Model for the Seismic Rehabilitation of Buildings*. Washington, DC: Federal Emergency Management Agency, Building Seismic Safety Council.

FEMA-350. 2000. *Recommended Seismic Design Criteria for New Steel Moment-Frame Buildings*. Washington, DC: Federal Emergency Management Agency.

Manyena, S.B. 2006. "The Concept of Resilience Revisited." *Disasters* 30, no. 4, pp. 434–450.

Mazzoni, S., F. McKenna, M.H. Scott, G.L. Fenves, and B. Jeremic. 2006. *Open System for Earthquake Engineering Simulation* (OpenSees). Berkeley, CA.

Mirfarhadi, S.A., and H.E. Estekanchi. (2020). Value-based Seismic Design of Structures Using Performance Assessment by the Endurance Time Method. *Structure and Infrastructure Engineering*, 16 no. 10, pp. 1397–1415.

Mirzaee, A., and H.E. Estekanchi. 2015. "Performance-Based Seismic Retrofitting of Steel Frames by Endurance Time Method." *Earthquake Spectra* 31, no. 1, pp. 383–402. http://dx.doi.org/10.1193/081312EQS262M.

Mirzaee, A., H.E. Estekanchi, and A. Vafai. 2010. "Application of Endurance Time Method in Performance-Based Design of Steel Moment Frames." *Scientia Iranica* 17, no. 6, pp. 361–70.

Mirzaee, A., H.E. Estekanchi, and A. Vafai. 2012. "Improved Methodology for Endurance Time Analysis: From Time to Seismic Hazard Return Period." *Scientia Iranica* 19, no. 5, pp. 1180–1187.

Mitropoulou, C.C., N.D. Lagaros, and M. Papadrakakis. 2010. "Building Design Based on Energy Dissipation: A Critical Assessment." *Bulletin of Earthquake Engineering* 8, no. 6, pp. 1375–1396.

Mitropoulou, C.C., N.D. Lagaros, and M. Papadrakakis. 2011. "Life-Cycle Cost Assessment of Optimally Designed Reinforced Concrete Buildings Under Seismic Actions." *Reliability Engineering & System Safety* 96, no. 10, pp. 1311–1131.

Wen, Y., and Y. Kang. 2001. "Minimum Building Life-Cycle Cost Design Criteria. II: Applications." *Journal of Structural Engineering* 127, no. 3, pp. 338–346.

Yang, T.Y., J.P. Moehle, and B. Stojadinovic. 2009. *Performance Evaluation of Innovative Steel Braced Frames*. Berkeley: University of California.

11 Sample Engineering Application

EDR Seismic Performance

11.1 INTRODUCTION

While the "smart structure" concept has been applied in aerospace and mechanical industries since a relatively long time, application of this concept for wind and seismic response reduction of civil engineering structures is still a cutting-edge technology under research and development (Cheng, Jiang, and Lou 2008).[1] A smart structure can be designed by applying various types of structural control devices. The four well-known types of these devices are seismic isolation, passive, semiactive and active, as well as hybrid control systems. Passive devices have the virtue of being more economical and less complicated compared to other types of control devices.

Friction dampers are among typical passive energy dissipating systems. Friction is an efficient, reliable, and economical mechanism that can dissipate kinetic energy by converting it to heat, so it can be used to slow down the motion of buildings. The function of friction devices in a building is analogous to the function of the braking system in an automobile (Soong and Dargush 1997). Based primarily on this analogy, Pall, Marsh, and Fazio (1980) began the development of friction dampers to improve the seismic response of civil engineering structures. Some of the most conventional types of these devices are the X-braced friction damper (Pall and Marsh 1982), Sumitomo friction damper (Aiken and Kelly 1990), Energy Dissipating Restraint (EDR) (Nims et al. 1993), and Slotted Bolted Connection (SBC) (FitzGerald et al. 1989), to name but a few.

The focus of this chapter is on application of the ET method in studying the performance of the EDR. EDR is a uniaxial friction damper designed by Richter and colleagues (1990). The mechanics of this device are described in detail in (Nims et al. 1993) and (Inaudi and Kelly 1996). The principal components of the device are internal spring, compression wedges, friction wedges, stops, and cylinder (Figure 11.1). The variable parameters are the number of wedges, spring constant, gap, and spring precompression. The role of the compression and friction wedges is to transmit and convert the axial force of the internal spring to a normal force on the cylinder wall.

The length of the spring can be variable through the operation of the device, which leads to a variable sliding friction. By adjusting the initial slip force and gap size,

DOI: 10.1201/9781003217473-11

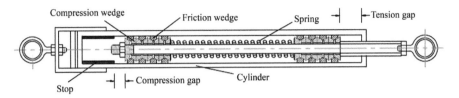

FIGURE 11.1   Configuration of the EDR.

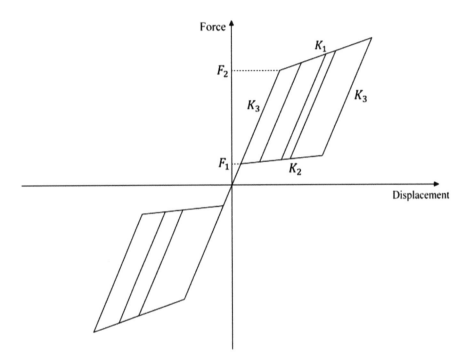

FIGURE 11.2   Double flag-shaped hysteretic loops of the EDR.

different hysteresis loops can be produced. With zero gaps and an initial slip force, the double flag-shaped loops result, as indicated in Figure 11.2. The parameters of the device are displayed in this figure. These double flag-shaped loops manifest the self-centering characteristic – that is, while unloading to zero, the device will return to its initial position without any residual deformations.

Several experimental studies have been carried out on the device, the results of which indicate the effectiveness of the EDR in reducing the seismic response of structures (e.g., Richter et al. 1990 and Aiken et al. 1993). The remarkable results are that the flag-shaped loops prove to be well-defined and quite consistent. Moreover, investigating cumulative energy time histories under earthquake signals implies that the frictional devices dissipate a significant portion of the total input energy.

The adequacy of the EDR has also been verified through manifold analytical studies (e.g., Inaudi and Kelly 1996; Inaudi and Nicos 1996; and Zhou and Peng

2009). However, as noted by the researchers, hysteretic mechanisms do not respond quickly to sudden impulses. Additionally, higher modes were sometimes excited due to sudden stiffness changes associated with the frictional devices. These limitations showed that the EDR device consistently provided reductions in displacements and interstory drifts, and increased the effective damping ratio of the test structure.

One of the most outstanding properties of the EDR, when the device is adjusted to have zero gaps, is that it is self-centering. As mentioned earlier, in this case, the EDR demonstrates double flag-shaped hysteresis loops. This property has the merit of reducing the permanent deformations in buildings after severe earthquakes. No other conventional friction damper enjoys this characteristic (Nims, Richter, and Bachman 1993). In fact, for the conventional friction dampers, which lack the self-centering property, significant permanent displacements could remain in the structure after the completion of the ground motion. This, in turn, brings about remarkable damage-repair costs. From now on, wherever the EDR is mentioned in this chapter, it refers to the device with double flag-shaped loops.

As a result of the highly nonlinear behavior of friction dampers, the use of the demanding nonlinear time-history method is inevitable for their reliable analysis and design. In fact, the alternate simplified methods that have been authorized by existing codes (e.g., nonlinear static and response spectrum procedures) are not reliable enough, on account of manifold simplifying postulations made in their development. Time history has the advantage of potentially being capable of directly including almost all sources of nonlinear and time-dependent material and geometric effects. Nevertheless, its traditional pitfall is being the most complex and time-consuming procedure.

As is apparent from Figure 11.2, the hysteretic behavior of the EDR device exhibits high nonlinearity. Therefore, it is necessary to apply the nonlinear time-history method for the performance-based analysis and design of the EDR-controlled structures. In the next section, application of "Endurance Time (ET)" method for analysis of these systems will be explained. The ET method is not as complicated and computationally demanding as the conventional time-history analysis. At the same time, it is not as unreliable and approximate as simplified methods. In fact, this method offers a more practical procedure for performance-based design of structures.

In this study, the application of the ET method in the performance-based seismic rehabilitation of the steel frames by using the EDR devices is investigated. Three steel moment-resisting frames with different story numbers are considered as case studies. By applying the ET method, the performance of the frames before and after installing the EDR devices is compared with each other. Several engineering demand parameters (including interstory drift, plastic rotation of beams and columns, and absolute acceleration) are employed to this end. Furthermore, the maximum interstory drift responses are also calculated through the nonlinear time-history analysis, using real ground motions, and the results are compared with those from the ET method.

11.2   BASIC CONCEPTS FROM THE ENDURANCE TIME METHOD

Among various conventional methods for the analysis of structures subjected to earthquake loadings, the nonlinear time-history analysis procedure is expected

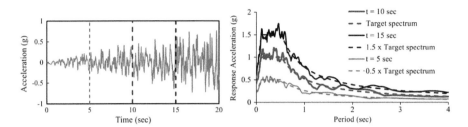

FIGURE 11.3 A typical ETEF and its response spectra at different times, along with the target spectra.

to produce the most realistic prediction of structural behavior. However, the complexity and high computational effort associated with this procedure have encouraged researchers to develop alternate analysis methods. These methods are much less complicated and can estimate the seismic demands to an acceptable degree of accuracy. The ET method is one of these methods that is applicable in many practical situations (Estekanchi et. al. 2020). This method is a time-history-based pushover procedure, in which the structure is subjected to a set of predesigned accelerograms that intensify with time – referred to as Endurance Time Excitation Functions (ETEFs). The ETEFs are generated in such a way that their response spectra increase according to the time; hence, the response of the structure under this kind of excitation gradually increases with time (Estekanchi, Valamanesh, and Vafai 2007). In other words, each time in an ETEF is a repre-sentative of a record with a certain level of intensity (Figure 11.3). In the process of generating excitation functions, the ETEFs are normally optimized to fit a spe-cific target spectrum, which could be a codified spectrum or the average spectrum of an ensemble of ground motions. In that process, the linear spectral acceleration of an ETEF is adjusted to satisfy Equations (11.1) and (11.2):

$$S_a\left(T,t\right)=\frac{t}{t_{target}}S_{aC}\left(T\right) \tag{11.1}$$

$$S_u\left(T,t\right)=\frac{t}{t_{target}}S_{uC}\left(T\right)\times\frac{T^2}{4\pi^2} \tag{11.2}$$

where T is the free vibration period of the Single-Degree-of-Freedom (SDOF) system; t is the time in the ETEF; S_a and S_u are the ETEFs' spectral acceleration and displacement spectra, respectively; S_{aC} and S_{uC} are the codified acceleration and displacement spectra, respectively; and t_{target} is a predefined time (equals 10 seconds) at which S_a and S_u coincide with S_{aC} and S_{uC}, respectively (Estekanchi, Valamanesh, and Vafai 2007). The performance of the structure is estimated based on the length of time during which it can endure the imposed ETEF. By using a properly designed excitation function, this endurance can be correlated to the intensity level of ground motions that the intended structure can be expected to endure. More description on the concept of the ET method as well as the characteristics of the ET excitation functions can be found in literature (e.g., Estekanchi, Vafai, and Sadeghazar 2004; Estekanchi, Valamanesh, and Vafai 2007; and Nozari and Estekanchi 2011, Mashayekhi et. al. 2018).

The main advantage of the ET method over the regular time-history method, using ground motions, is that it needs a small number of analyses. In the ET method, the structural responses at different excitation levels are obtained in a single time-history analysis, thereby significantly reducing the computational demand. Accordingly, by using the ET method and regarding the concepts of performance-based design, the performance of a structure at various seismic hazard levels can be predicted in a single time-history analysis. The application of the ET method in the seismic performance assessment of steel frames has been studied by Mirzaee and Estekanchi (2015), and Mirfarhadi and Estekanchi (2020).

The results of ET analysis are usually presented by increasing ET response curves. The ordinate at each time value, t, corresponds to the maximum absolute value of the required engineering demand parameter in the time interval $[0, t]$, as is expressed in Equation (11.3):

$$\Omega\big(P(t)\big) \equiv \max(|P(\tau)|) \qquad \tau \in [0,t] \tag{11.3}$$

In this equation, Ω is the *Max-Abs* operator, as was defined above, and $P(t)$ is the desired response history such as interstory drift ratio, base shear, or other parameters of interest. The abscissa of an ET response curve is the analysis time, which is an indicator of the intensity in ET analysis. Figure 11.4(a) shows a typical ET response curve in which the maximum interstory drift is utilized as the demand parameter. ET curves are usually serrated, because of the statistical characteristics and dispersion of the results of the ET analysis in the nonlinear range. Sometimes the response value does not pass the maximum value experienced before in a time interval, and therefore the resulting ET curve has a constant value in that interval. In order to become more accurate and consistent ET curves, Estekanchi and colleagues recommended

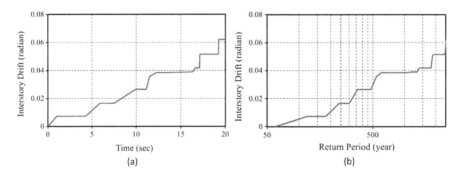

FIGURE 11.4   Sample ET response curve with horizontal axis in (a) time, (b) return period.

using the average of the results from three ET excitation functions (Estekanchi, Valamanesh, and Vafai 2007).

Mirzaee, Estekanchi, and Vafai (2012) originally investigated the correlation between time – as an indicator of the intensity in the ET analysis – and seismic hazard return period. Substituting a common parameter, such as the return period for time, increases the readability and efficiency of response curves and can considerably improve the presentation of ET analysis results. They utilized the elastic response spectrum defined in ASCE (2006) as an intermediate criterion to establish this correlation in their presentation.

Further investigations suggested that utilizing the elastic spectrum as the intermediate intensity measure to correlate the time and return period is not the best choice in the cases in which the structures experience large nonlinear deformations (Estekanchi, Riahi, and Vafai 2011; Riahi, Estekanchi, and Vafai 2009). Actually, in the structures that experience large nonlinear deformations, the difference between the results obtained by this procedure and the nonlinear time-history analysis, using ground motions, can be significant.

Foyouzat and Estekanchi (2014) proposed using nonlinear Rigid-Perfectly Plastic (RPP) spectra in lieu of elastic response spectra to correlate the time in the ET analysis and return period. The results suggested that the application of RPP spectra significantly improves the accuracy and reliability of the response curves that result from ET analysis in nonlinear range compared with the procedures based on linear elastic spectra. As a result, regarding the high nonlinearity associated with the EDR device, as is discussed in the previous section, the RPP spectra are more appropriate intensity measures than the elastic spectra for the ET analysis of the structures outfitted with EDR devices. In what follows, a brief explanation of this approach, which is discussed in detail in Foyouzat and Estekanchi (2014), is presented. An RPP system possesses a force-displacement relationship, as indicated in Figure 11.5. No deformation occurs until F reaches the yield force, F_y, and the force cannot exceed the yield force, that is, $|F| \leq F_y$. The RPP model can be simulated by a Coulomb friction block with a sliding friction force equal to F_y.

For a given earthquake excitation, the response of an RPP SDOF system depends only on the ratio $A_y = F_y/m$, where m is the mass of the SDOF system. For a given ground motion, if maximum absolute displacements of RPP SDOF systems are calculated for a range of A_ys, the RPP spectrum of that ground motion will be obtained. Furthermore, if the ground motion is scaled to a seismic hazard level corresponding to a specific return period, the resulting spectrum is the RPP spectrum corresponding to that return period. Besides this, one could obtain the RPP spectra of an ensemble of ground motions that are scaled to a specific return period and then use the average of those spectra as the RPP spectrum corresponding to that return period.

As is clear from the discussion above, the RPP spectrum is a function of two variables, namely the return period (R) and A_y/g, that is to say $S_{RPP} = S_{RPP} (A_y/g, R)$. Apart from this, the RPP spectrum for an ETEF is defined as is indicated in Equation (11.4):

$$S_{RPP}\left(A_y / g,t\right) = \max(|\Delta(\tau)|) \qquad \tau \in [0,t] \qquad (11.4)$$

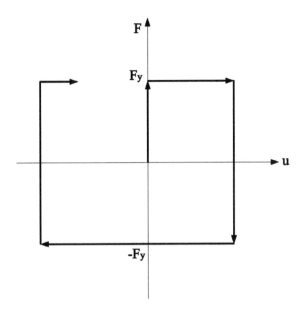

FIGURE 11.5   Force-displacement behavior of an RPP model.

where t is time, and $\Delta(\tau)$ is the displacement of the RPP system at time A due to an ETEF. If more than one ETEF is to be used (usually three, as pointed out earlier), the average spectra of those ETEFs can be applied. By acquiring the inverse of the function S_{RPP} with respect to R, the return period can be written as $R = f(S_{RPP}, A_y/g)$, where f is a function that relates the equivalent earthquake return period to S_{RPP} and A_y/g. From Equation (11.4), $S_{RPP} = S_{RPP}(A_y/g, t)$, the result of which would be Equation (11.5):

$$R = f\left(S_{RPP}\left(A_y/g,t\right), A_y/g\right) = h\left(A_y/g,t\right) \tag{11.5}$$

where h is a function that relates the return period to A_y/g and t. Since expressing function h via a closed form formulation is difficult, this function is evaluated numerically in a range of A_y/gs and ts, and the values of R can be stored in a matrix format as done by Foyouzat and Estekanchi (2014). The ET time at which Equation (11.5) holds is referred to as the *equivalent time* corresponding to return period R and A_y/g.

In order to calculate the parameter A_y/g of a structure, it is usually adequate to use the pushover curve resulting from a load pattern that is based on the first elastic mode shape. The effective yield force that is obtained from the pushover curve is divided by the mass of the structure to give parameter A_y. Having the return period and parameter A_y/g of the structure, one can, using Equation (11.5), readily get the ET equivalent time sought. After imposing an ETEF to the structure, the maximum absolute value of a desired response up to the equivalent time is calculated. If a set of ETEFs is considered, the average value is recorded as the response demand corresponding to

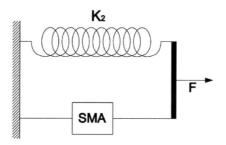

FIGURE 11.6   Modeling the EDR behavior in OpenSees.

the considered return period. This process is renewed for several return periods until the response curve of the structure is acquired. A typical response curve produced in this way is shown in Figure 11.4(b). The return period axis is plotted in a logarithmic scale.

11.3   MODELING THE EDR DEVICE IN OPENSEES

OpenSees is one of the best software for modeling highly nonlinear macro-modeling problems. Thus, in this study, all nonlinear analyses are performed in OpenSees (PEERC 2013). Unfortunately, in OpenSees there are no predefined materials that behave like EDR in loading and unloading phases. However, by assembling a few uniaxial materials, the EDR behavior can be modeled easily.

Let us consider the flag-shaped loops whose parameters are indicated in Figure 11.2. In order to model the flag-shaped behavior in OpenSees, one can combine a uniaxial element with linear elastic behavior and a uniaxial element with Shape Memory Alloy (SMA) behavior in parallel (Figure 11.6). The hysteretic behavior of an SMA uniaxial element in OpenSees, together with the required parameters, is displayed in Figure 11.7. If the stiffness of the linear elastic element is set equal to K_2, the parameters of the SMA are given by Equations (11.6) through (11.9):

$$E = K_3 - K_2 \tag{11.6}$$

$$\sigma_s^{AM} = \frac{\left(K_3 - K_2\right)F_2}{K_3} \tag{11.7}$$

$$\sigma_s^{MA} = \sigma_f^{MA} = \frac{\left(K_3 - K_2\right)F_1}{K_3} \tag{11.8}$$

$$\sigma_f^{AM} = \frac{\left(K_3 - K_2\right)F_2}{K_3} + \frac{\left(K_1 - K_2\right)\left(K_3 - K_2\right)\varepsilon_L}{K_3 - K_1} \tag{11.9}$$

Since the hardening part of the SMA is not present in the EDR loops, the magnitude of ε_L can be taken a very large value so this point that actually can never be reached in the numerical analyses. It has been observed that F_1 is not independent of other four parameters (Inaudi and Kelly 1996). In fact, it can be shown that Equation (11.10) holds between these five parameters of the EDR:

$$F_1 = \frac{K_2\left(K_3 - K_1\right)}{K_1\left(K_3 - K_2\right)} F_2 \qquad (11.10)$$

As a result, only four parameters are needed so as to completely model the EDR device.

11.4 PERFORMANCE ASSESSMENT OF EDR DEVICES BY ET METHOD

In this section, the effectiveness of EDR devices in controlling the seismic response of structures is investigated by applying the ET analysis method. Three steel moment resisting frames (MRFs) with different numbers of stories are addressed as the case studies. The set under investigation consists of two-dimensional frames, with 3, 6, and 10 stories and 3 bays, and built on a site in the Los Angeles region with Soil Class C, as defined by ASCE (American Society of Civil Engineers 2006). The height of all stories is 3.2m, and the bay width is 5.0m. Some basic properties of these frames are summarized in Table 11.1. In this table, parameter A_y/g is calculated according to the procedure explained in Section 11.2.

The supports of the 3st3bINITIAL and 6st3bINITIAL frames are assumed to be fixed, while hinged supports are chosen for the 10st3bINITIAL frame. The first story of frame 6st3bINITIAL is assumed to be surrounded by a concrete retaining wall, which forces it to experience similar lateral displacements to the ground at any time. As a result, the base level of this frame is transferred to the level of the first story. These structures are designed by applying only a fraction of the codified design base shear per INBC code (Iranian National Building Code 2005) – which is quite consistent with the AISC-ASD building code (American Institute of Steel Construction 1989) – so that the structures will require rehabilitation by using EDR devices. Additionally, it is assumed that, for practical reasons, the owner has constrained the installation of the EDRs to only one bay in each story.

TABLE 11.1
Properties of the Initial Frames in Summary

Property	3st3bINITIAL	6st3bINITIAL	10st3bINITIAL
Mass participation (mode 1)	81%	77%	78%
Fundamental period, T_1 (sec)	0.97	1.24	1.6
A_y/g	0.24	0.17	0.23

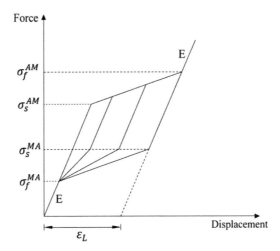

FIGURE 11.7   The hysteretic behavior of an SMA uniaxial element in OpenSees.

TABLE 11.2
Description of the Ground Motions Used in This Study

ID No.	Year	Earthquake name	Station	Component (deg)	PGA (m/s²)
1	1999	Kocaeli, Turkey	Arcelik	0	2.15
2	1976	Friuli, Italy	Tolmezzo	0	3.45
3	1995	Kobe, Japan	Nishi-Akashi	0	5.00
4	1999	Hector Mine	Hector	90	3.30
5	1986	Palmsprings	Fun Valley	45	1.29
6	1979	Imperial Valley	El Centro, Parachute Test	315	2.00
7	1984	Morgan Hill	Gilroy #6, San Ysidro	90	2.80

Table 11.2 describes the ground motions employed for verification in the current study. All of these ground motions are recorded on Soil Type C. The scale factor for each ground motion, corresponding to return period R, is selected so that the 5 percent damping linear elastic spectrum of the ground motion between $0.2T$ and $1.5T$ will not fall below the codified spectrum (corresponding to return period R) in the same range, where T is the fundamental period of the structure being analyzed. The codified spectrum corresponding to any return period is formulated in ASCE41. After the calculation of the scale factors, the average RPP spectrum corresponding to each return period can be obtained. For example, the RPP spectra for the ensemble of ground motions scaled to the return period of 475-yr for $T = 1$ sec, together with their average, are represented in Figure 11.8 (g is the acceleration of gravity).

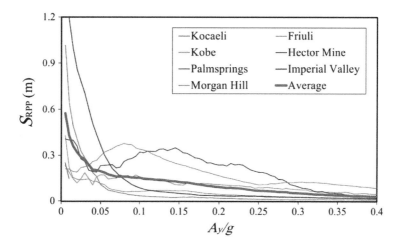

FIGURE 11.8   The RPP spectra of an ensemble of scaled ground motions together with their average.

The ETA20inx01-3 series, generated with the duration of 20 seconds, is used as ETEFs. More information on different ETEF series is publicly available on the ET method website (Endurance Time Method website 2021). The RPP spectrum for each ETEF can be acquired by applying Equation (11.4). Equivalent times can now be calculated through using Equation (11.5) and, last, by pursuing the procedure explained in Section 11.2, the ET response curves can be achieved. The provisions of ASCE41-06 are applied for performance-based design and check of the structures. The rehabilitation objective is selected as Enhanced Objectives – k and p and b, as defined in ASCE41-06.

Before getting down to the analysis of the three aforementioned frames, it is worth introducing two useful definitions originally proposed by Mirzaee and Estekanchi (2015). These definitions can facilitate the evaluation of the seismic performance of the structures using the ET method. The first definition is referred to as the Damage Level (DL) index, which is a normalized continuous numerical value as defined by Equation (11.11). The DL is a dimensionless index that creates a numerical presentation for performance levels – values of 1, 2, and 3 for IO (Immediate Occupancy), LS (Life Safe), and CP (Collapse Prevention) levels, respectively).

$$DL = \sum_{i=1}^{n} \frac{\max[\theta_{i-1}, \min(\theta, \theta_i)] - \theta_{i-1}}{\theta_i - \theta_{i-1}} \qquad (11.11)$$

In this equation, θ is the parameter that should be computed from the analyses and checked as per codes in order to evaluate the seismic behavior of the structure. The parameter θ can be a representative of the plastic rotation in beams, the plastic rotation in columns, or any other significant response parameter for which limiting values as per codes have been adopted. Additionally, n is the number of performance levels considered in the design ($n = 3$ in this study). The parameters θ_i are the ASCE41-06 limiting values at each performance level, and θ_0 is always set equal to zero. It should be noted that the DL index is not a new response parameter in addition to those addressed in ASCE41 for the evaluation of the structures. It is only a new form of representing the responses on a normalized continuous numerical scale. Moreover, utilizing the DL index facilitates the combination of different parameters that are involved in assessing the seismic performance of a structure. The second definition is referred to as the *target curve*. The target curve specifies maximum acceptable responses at various DLs as a continuous curve (Mirzaee and Estekanchi 2015). By comparing the ET performance curve with the target curve, the seismic performance of the structure at different seismic intensities can be evaluated.

Figures 11.9 through 11.12 illustrate the interstory drift, plastic rotation of columns, plastic rotation of beams, and absolute acceleration response curves for the foregoing frames, respectively. According to the ASCE41 provisions, the limiting values for the plastic rotation of beams depend on the section properties while, for the plastic rotation of columns, they depend on both the section properties and the axial force of the columns. Hence, the plastic rotations are represented in terms of the DL index in order to avoid multiple target curves and streamline the presentation of the diagrams. Since no acceptance criterion for the absolute acceleration has yet been established in ASCE41, the target curve is absent from the absolute acceleration diagrams. For the 10-story frame, the response curves are shown up to the return period of 1,500 years. The reason is that the duration of ETA20inx01-3 series (20 seconds) is not sufficient to cover all the return periods of interest. Generating ETEFs with longer durations can resolve this issue.

According to ASCE41-06, if the axial force to P_{CL} (the lower bound axial column strength) ratio of a column falls below 0.5, only the column rotation needs to be checked, and there is no need to check the axial force-bending moment interaction equation. As was observed in all the results of this study, in no column did this ratio exceed 0.5; thus, checking the interaction equation is no longer necessary. It is discernible from the response curves that the drifts and column rotations exceed the target curve in some cases and the structures need to be rehabilitated. To this end, EDR dampers are to be employed to control the seismic response of the structures. In the middle bay of each story, two identical EDR devices are installed in the form of cross bracings. The properties of each device are selected by trial and error until an acceptable response is achieved.

At each stage of the trial-and-error process, the fundamental period and parameter A_y/g of the rehabilitated structure are calculated. By doing so, the equivalent times are obtained, and which must be employed to acquire the ET response curves, as was previously explained. The A_y/g parameter does not significantly vary through stages, thanks to the relatively low stiffness of EDR devices. Therefore, the equivalent times undergo trivial changes at each stage as compared to the preceding stages. As a result, one can use the same A_y/g of the initial structure for the ensuing stages to avoid performing a separate pushover analysis for each stage. After reaching an acceptable response, a pushover analysis can be performed to obtain the exact A_y/g of

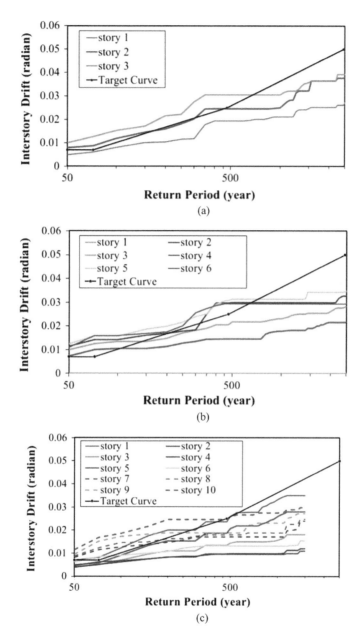

FIGURE 11.9   Interstory drift response curves for (a) 3st3bINITIAL, (b) 6st3bINITIAL, and (c) 10st3bINITIAL frames.

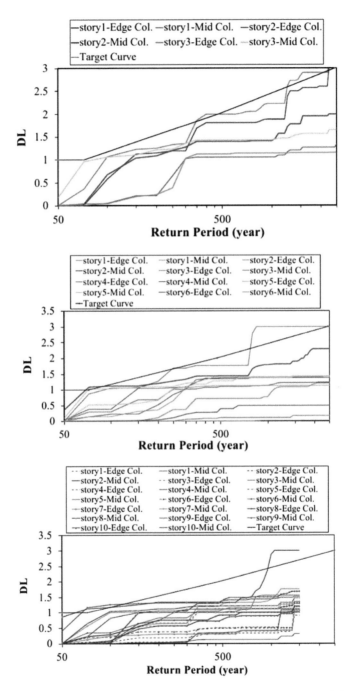

FIGURE 11.10 Column rotation response curves for (a) 3st3bINITIAL, (b) 6st3bINITIAL, and (c) 10st3bINITIAL frames.

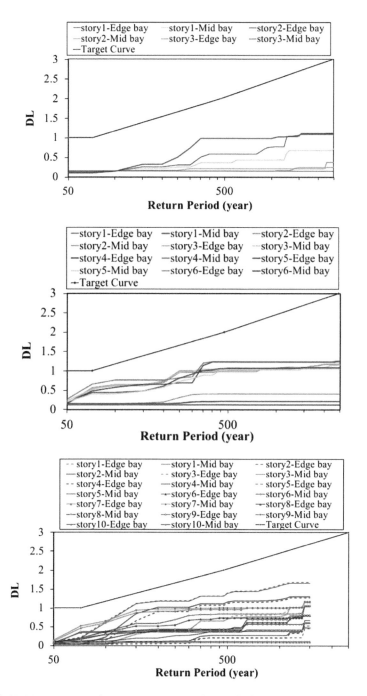

FIGURE 11.11 Beam rotation response curves for (a) 3st3bINITIAL, (b) 6st3bINITIAL, and (c) 10st3bINITIAL frames.

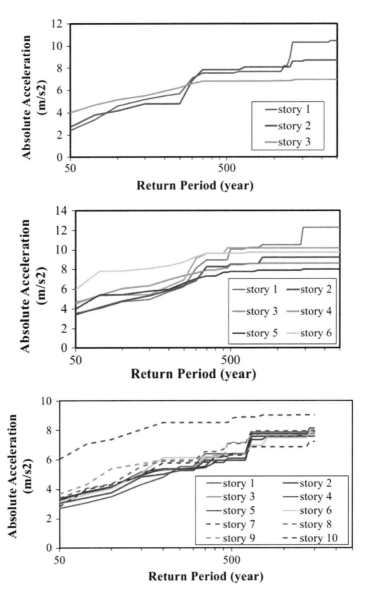

FIGURE 11.12 Absolute acceleration response curves for (a) 3st3bINITIAL, (b) 6st3bINITIAL, and (c) 10st3bINITIAL frames.

the last stage and reproduce the response curves. This procedure effectively reduces the design effort.

The characteristics of each device at the end of the trial-and-error process together with the parameters of the rehabilitated frames – referred to as 3st3bEDR, 6st3bEDR, and 10st3bEDR – are summarized in Table 11.3. The resulting response curves of the rehabilitated frames are shown in Figures 11.13 through 11.16. As can be inferred from these figures, EDR devices have significantly reduced the drifts and column

TABLE 11.3
Properties of the Rehabilitated Frames in Summary

Frame name	EDR location	K_1 (kN/m)	K_2 (kN/m)	K_3 (kN/m)	F_2 (kN)
3st3bEDR, T1 = 0.85	1st story	1,339	17.8	13,388	60
sec, Ay/g = 0.26	2nd story	1,339	17.8	13,388	60
	3rd story	964	17.8	9,640	40.1
6st3bEDR, T1 = 1.13	2nd story	1,607	17.8	16,066	60
sec, Ay/g = 0.19	3rd story	1,339	17.8	13,388	60
	4th story	1,607	17.8	16,066	60
	5th story	1,071	17.8	10,711	60
	6th story	428	17.8	4,284	40.1
10st3bEDR,	1st story	4,820	17.8	48,198	50
T1 = 1.51 sec,	2nd story	2,142	17.8	21,421	50
Ay/g = 0.24	3rd story	2,410	17.8	24,099	50
	4th story	1,500	17.8	14,995	40
	5th story	1,071	17.8	10,711	40
	6th story	1,071	17.8	10,711	40
	7th story	1,285	17.8	12,853	50
	8th story	1,285	17.8	12,853	50
	9th story	1,285	17.8	12,853	50
	10th story	1,285	17.8	12,853	50

rotations of the structures in large and medium return periods, which correspond to moderate and strong ground motions, respectively.

Despite this issue, there is only a slight reduction in lower return periods, and the interstory drift response curves do not completely fall below the target curve in this range. The main reason is that, in small events, few hysteresis loops develop, and a small amount of energy is dissipated. However, in medium and large return periods, the formation of quite a few loops dissipates a large amount of energy, which causes the responses to be considerably mitigated. Even by increasing the slip forces (F_2s) of the devices, the responses do not improve effectively in small return periods. Similarly, the use of a higher initial stiffness is not useful, since this will increase the axial force demand of the damper, which causes the device to fail.

It is worth noting that the interstory drift acceptance criteria stipulated in ASCE41-06, are only recommended values, and it is not imperative for a structure to satisfy them. In fact, if a design can meet the beam and column plastic rotation acceptance criteria (and, if necessary, the axial force-bending moment interaction effect), it is rated as an acceptable design. Accordingly, the designs of the rehabilitated frames are acceptable regarding the performance objectives.

Referring to the foregoing results, it can be concluded that if it is desired to provide additional damping for a range of moderate and large earthquakes, the EDR device is an apt choice. This would be the case, provided that the building performance for small events is satisfactory, and also limiting the device force is important. Through applying the time-history analysis method on a range of SDOF structures, Nims and colleagues (1993)

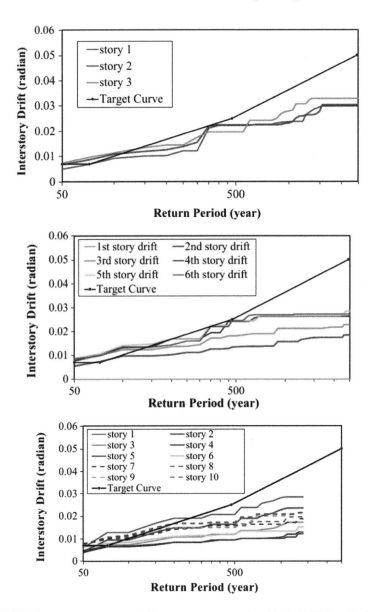

FIGURE 11.13 Interstory drift response curves for (a) 3st3bEDR, (b) 6st3bEDR, and (c) 10st3bEDR frames.

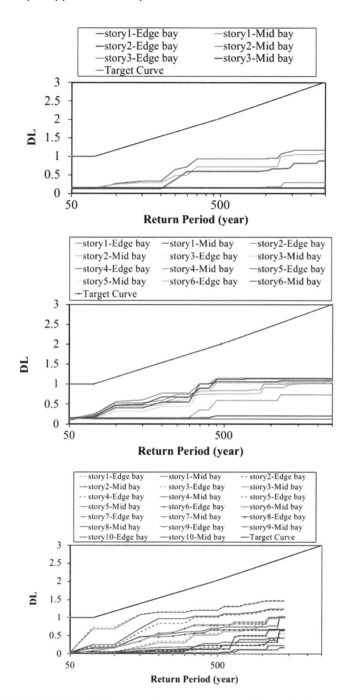

FIGURE 11.14 Column rotation response curves for (a) 3st3bEDR, (b) 6st3bEDR, and (c) 10st3bEDR frames.

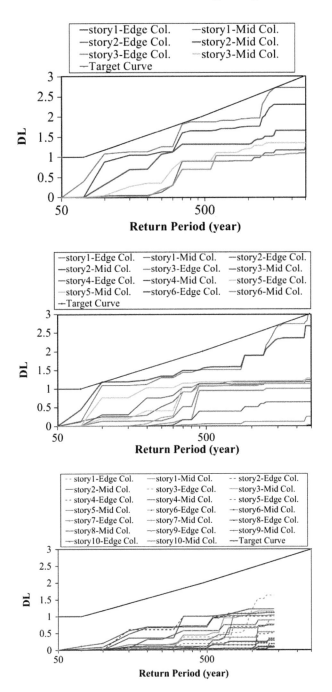

FIGURE 11.15 Beam rotation response curves for (a) 3st3bEDR, (b) 6st3bEDR, and (c) 10st3bEDR frames.

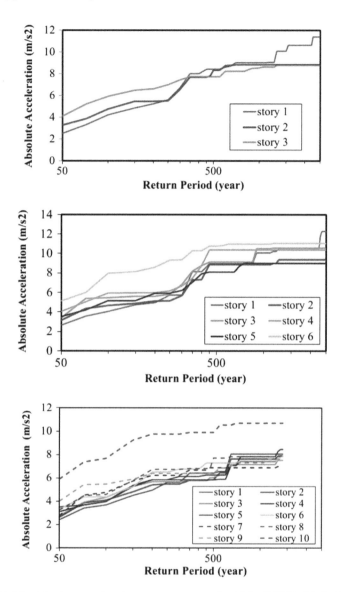

FIGURE 11.16 Absolute acceleration response curves for (a) 3st3bEDR, (b) 6st3bEDR, and (c) 10st3bEDR frames.

drew a similar conclusion for SDOF systems. By using a more affordable ET method, the current study has verified this result for real multistory frames.

Figures 11.12 and 11.16 display (before and after the rehabilitation, respectively) the absolute acceleration response curves of the foregoing frames. The absolute acceleration is among those parameters that play an important role in the nonstructural damage, the life-cycle cost due to the loss of contents (Elenas and Meskouris 2001), and the occupants' comfort (Connor 2002). Generally, moment-resisting frames have acceptable values for absolute acceleration, while the absolute accelerations in braced frames are large. As can be observed from Figures 11.12 and 11.16, compared to that of the initial structures the installation of the EDR devices have not significantly increased the absolute accelerations. This result reveals one of the advantages of these devices. The reason for this behavior lies in the relatively small stiffness of the EDRs as well as the energy absorption due to their operation during the earthquake.

As stated previously, as far as the absolute acceleration is concerned, no acceptance criterion has been stipulated in ASCE41-06. However, several researchers have proposed limiting values for the absolute acceleration at different performance levels. For example, according to Elenas and Meskouris (2001), for the IO, LS, and CP levels, the corresponding limiting values are 2.0, 9.8, and 12.5 m/s^2, respectively. Figure 11.16 suggests that, except for the small return periods, the absolute accelerations satisfy the above limitations.

11.5 COMPARATIVE STUDY

In this section, a comparative study is carried out between three different methods of analysis, namely time-history analysis, the ET method based on RPP spectra, and the ET method based on elastic spectra. The results of the second method were obtained in the previous section. The last method, as was discussed in preceding sections, utilizes the elastic spectra as the intermediate intensity measure to correlate the time and return period. A detailed description of this method can be found in the study by Mirzaee, Estekanchi, and Vafai (2012). Apart from this, the time-history analysis is performed by using the ground motions described in Table 11.2. The maximum interstory drift responses of the aforementioned frames are calculated via these three methods in a number of return periods, some results of which are displayed in Figure 11.17. Note that the time-history responses in Figure 11.17 are based on the average of the maximum absolute values resulting from the analyses over the considered ground motions.

Figure 11.17 suggests that in medium and large return periods (i.e., return periods greater than 475 years), the results of the ET method based on RPP spectra show good concordance with the results of the time-history analysis. In addition, the trends of the diagrams are well predicted by the ET method. This stems from the fact that the frames experience significant inelastic displacements in these return periods. However, in small return periods, the frames experience slight plastic deformations. Therefore, in small return periods the ET method that is based on elastic spectra yield a better result, although the RPP spectra-based method is still a good approximation.

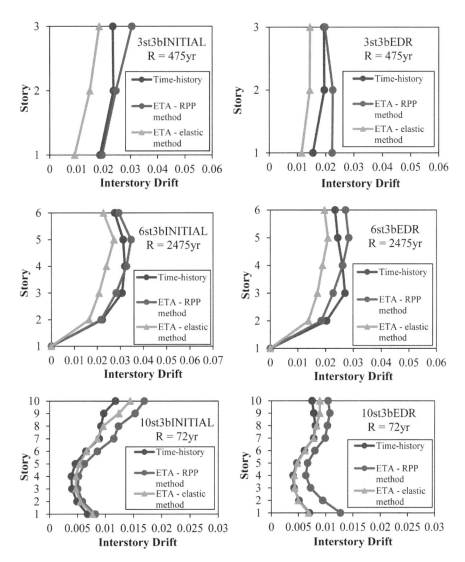

FIGURE 11.17 Comparison of the maximum interstory drift responses of the frames under study for a number of return periods calculated via time-history analysis, ET method based on RPP spectra and ET method based on elastic spectra.

Basically, these observations are in line with the results reported in Foyouzat and Estekanchi (2014).

It must be noted that the RPP model has zero plastic phase slope, while the slope of the second portion of the EDR devices is a nonzero value (see Figure 11.2). Additionally, the hysteresis loops of the EDR devices are completely different from the loops of the RPP materials. Apart from this, the inherent dispersion of the results of the ETEFs used in the ET method can be another important source of error. In spite of these important differences, the results of the ET method based on RPP spectra, as previously observed, show an acceptable degree of accuracy. Moreover, the ET results are conservative in quite a few cases.

Another important point is that, in the time-history method, the structural responses were generated by using seven ground motions at three return periods, requiring 21 time-history analyses. On the other hand, generating the responses by the ET method required only three time-history analyses. Furthermore, if it is needed to calculate the responses in more return periods, the required number of the analyses increases proportionally in the time-history method whereas, when the ET method is utilized, it remains the same (i.e., three analyses).

11.6 SUMMARY AND CONCLUSIONS

In this chapter, the application of the ET method in the performance assessment of EDR friction devices for the seismic rehabilitation of steel frames was investigated. Three steel MRFs with different numbers of stories are considered as the case studies. Double flag-shaped EDR devices are employed in order to improve the seismic response of the initial frames. The behavior of these dampers is highly nonlinear in comparison with other friction dampers. Accordingly, the improved ET method, which is based on nonlinear RPP spectra, is applied in order to satisfactorily estimate the responses in nonlinear range. From the results of this study, the following conclusions can be drawn:

1. In medium and large return periods, double flag-shaped EDR devices effectively improve the seismic responses of the initial frames and reduce the demand parameters to acceptable codified values.
2. In small return periods, the seismic responses of the frames are not considerably improved. Therefore, if the initial frame significantly fails to satisfy performance limits corresponding to small return periods (for instance, the 72-year return period), double flag-shaped EDRs cannot effectively mitigate the responses to reach the allowable limits.
3. As a result of the relatively small stiffness of the EDRs as well as the energy absorption due to their operation during the earthquake, the installation of the EDR devices does not significantly increase the absolute acceleration of stories compared to that of the initial structures.
4. The application of the RPP spectra improves the accuracy and reliability of the response curves resulted from ET analysis in nonlinear range compared with the procedures based on linear elastic spectra. The results of the RPP spectra-based method show good concordance with the results of the time-history analysis.

As far as the computational cost is concerned, the ET method is far more economical in comparison with the conventional time-history method. Moreover, the ET method enjoys high reliability and accuracy compared to the alternate simplified methods and enables the evaluation of the seismic performance as a continuous function of seismic hazard return period. As a result, the ET method can be effectively employed for the multilevel performance-based seismic rehabilitation of structures.

Since friction is an effective, reliable, and economical mechanism that can dissipate the energy introduced to structures by seismic events, the use of this mechanism can be highly desirable in seismic rehabilitation. The EDR is a self-centering friction device; thus, it can alleviate the permanent deformations of structures after the completion of the earthquake, leading to decreased damage repair costs. The hysteretic behavior of this device is highly nonlinear, so the use of the demanding nonlinear time-history analysis is a requisite for the frames whose responses have been controlled through EDR devices. Applying the ET method can surmount the intricacy of the time-consuming nonlinear time-history analysis and put a more practical and favorable way at the disposal of structural designers in order to exploit the EDR friction mechanism for seismic hazard mitigation.

NOTE

1 *Chapter Source*: Foyuzat, M.A., and H.E. Estekanchi. 2016. "Evaluation of the EDR Performance in Seismic Control of Steel Structures Using Endurance Time Method." *Scientica Iranica* 23, no. 3, pp. 827–841.

REFERENCES

Aiken, I.D., and J.M. Kelly. 1990. *Earthquake Simulator Testing and Analytical Studies of Two Energy-absorbing Systems for Multistory Structures*. Report No. UCB/EERC-90/ 03. Earthquake Engineering Research Center, College of Engineering, University of California, Berkeley.

Aiken, I.D., D.K. Nims, A.S. Whittaker, and J.M. Kelly. 1993. "Testing of Passive Energy Dissipation Systems." *Earthquake Spectr* 9, no. 3, pp. 335–370.

American Institute of Steel Construction (AISC). 1989. *Manual of Steel Construction: Allowable Stress Design*, 9th ed. Chicago.

American Society of Civil Engineers (ASCE). 2006. *Seismic Rehabilitation of Existing Buildings*, *ASCE/SEI 41-06*. Reston, VA.

Cheng, F.Y., H. Jiang, and K. Lou. 2008. *Smart Structures: Innovative Systems for Seismic Response Control*. Boca Raton, Florida: CRC Press, Taylor & Francis Group.

Connor, J.J. 2002. *Introduction to Structural Motion Control*. MIT/Prentice Hall.

Elenas, A., and K. Meskouris. 2001. "Correlation Study Between Seismic Acceleration Parameters and Damage Indices of Structures." *Engineering Structures* 23, pp. 698–704.

Endurance Time Method website. 2021. https://sites.google.com/site/etmethod/.

Estekanchi, H.E., M. Mashayekhi, H. Vafai, G. Ahmadi, S.A. Mirfarhadi, and M. Harati. (2020). "A State-of-knowledge Review on the Endurance Time Method." *Structures* v27, pp. 2288–2299.

Estekanchi, H.E., H.T. Riahi, and A. Vafai. 2011. "Application of Endurance Time Method in Seismic Assessment of Steel Frames." *Engineering Structures* 33, pp. 2535–2546.

Estekanchi, H.E., A. Vafai, and M. Sadeghazar. 2004. "Endurance Time Method for Seismic Analysis and Design of Structures." *Scientia Iranica* 11, no. 4, pp. 361–370.

Estekanchi, H.E., V. Valamanesh, and A. Vafai. 2007. "Application of Endurance Time Method in Linear Seismic Analysis." *Engineering Structures* 29, no. 10, pp. 2551–2562.

FitzGerald, T.F., T. Anagnos, M. Goodson, and T. Zsutty. 1989. "Slotted Bolted Connections in Aseismic Design for Concentrically Braced Connections." *Earthquake Spectra* 5, no. 2, pp. 383–391.

Foyouzat, M.A., and H.E. Estekanchi. 2014. "Application of Rigid-perfectly Plastic Spectra in Improved Seismic Response Assessment by Endurance Time Method." *Engineering Structures* 111, pp. 24–35.

Inaudi, J.A., and J.M. Kelly. 1996. "Dynamics of Homogeneous Frictional Systems." *Dynamics with Friction: Modeling, Analysis, and Experiments* 7, pp. 93–136.

Inaudi, J.A., and M. Nicos. 1996. "Time-domain Analysis of Linear Hysteretic Damping." *Earthquake Engineering and Structural Dynamics* 25, pp. 529–545.

Iranian National Building Code (INBC). 2005. *Design and Construction of Steel Structures.* Office of Collection and Extension of National Building Code, Ministry of Housing and Urban Development. Tehran.

Mashayekhi, M., H.E. Estekanchi, H. Vafai, and S.A. Mirfarhadi. 2018. "Development of Hysteretic Energy Compatible Endurance Time Excitations and Its Application." *Engineering Structures*, v177, pp. 753–769.

Mirfarhadi, S.A., and H.E. Estekanchi. (2020). "Value-based Seismic Design of Structures Using Performance Assessment by the Endurance Time Method." *Structure and Infrastructure Engineering*, 16, no. 10, pp. 1397–1415.

Mirzaee, A., and H.E. Estekanchi. 2015. "Performance-based Seismic Retrofitting of Steel Frames by Endurance Time Method." *Earthquake Spectra* 31, no. 1, pp. 383–402.

Mirzaee, A., H.E. Estekanchi, and A. Vafai. 2012. "Improved Methodology for Endurance Time Analysis: From Time to Seismic Hazard Return Period." *Scientia Iranica* 19, no. 5, pp. 1180–1187.

Nims, D.K., J.A. Inaudi, P.J. Richter, and J.M. Kelly. 1993. "Application of the Energy Dissipating Restraint to Buildings." *Proceedings of ATC 17-1 on Seismic Isolation, Passive Energy Dissipation, and Active Control* 2, pp. 627–638.

Nims, D.K., P.J. Richter, and R.E. Bachman. 1993. "The Use of the Energy Dissipating Restraint for Seismic Hazard Mitigation." *Earthquake Spectra* 9, no. 3, pp. 467–489.

Nozari, A., and H.E. Estekanchi. 2011. "Optimization of Endurance Time Acceleration Functions for Seismic Assessment of Structures." *International Journal of Optimization in Civil Engineering* 1, no. 2, pp. 257–277.

Pacific Earthquake Engineering Research Center (PEERC). 2013. *Open System for Earthquake Engineering Simulation.* (OpenSees), Berkeley: University of California. Available online http://opensees.berkeley.edu/.

Pall, A.S., and C. Marsh. 1982. "Response of Friction Damped Braced Frames." *Journal of Structural Division, ASCE* 108, no. 9, pp. 1313–1323.

Pall, A.S., C. Marsh, and P. Fazio. 1980. "Friction Joints for Seismic Control of Large Panel Structures." *Journal of Prestressed Concrete Institute* 25, no. 6, pp. 38–6l.

Riahi, H.T., H.E. Estekanchi, and A. Vafai. 2009. "Estimates of Average Inelastic Deformation Demands for Regular Steel Frames by the Endurance Time Method." *Scientia Iranica* 16, no. 5, pp. 388–402.

Richter, P.J., D.K. Nims, J.M. Kelly, and R.M. Kallenbach. 1990. "The EDR-energy Dissipating Restraint, a New Device for Mitigating Seismic Effects." *Proceedings of the 1990 SEAOC Convention* no. 1, pp. 377–401. Lake Tahoe.

Soong, T.T., and G.F. Dargush. 1997. *Passive Energy Dissipation Systems in Structural Engineering.* New York: John Wiley.

Zhou, X., and L. Peng. 2009. "A New Type of Damper with Friction-variable Characteristics." *Earthquake Engineering and Engineering Vibration* 8, no. 4, pp. 507–520.

Index